Survey of Electronics

Merrill's International Series in Electrical and Electronics Technology

BATESON	*Introduction to Control System Technology, 2nd Edition, 8255-2*
BOYLESTAD	*Introductory Circuit Analysis, 4th Edition, 9938-2*
	Student Guide to Accompany Introductory Circuit Analysis, 4th Edition, 9856-4
BOYLESTAD/KOUSOUROU	*Experiments in Circuit Analysis, 4th Edition, 9858-0*
BREY	*Microprocessor/Hardware Interfacing and Applications, 20158-6*
FLOYD	*Digital Fundamentals, 2nd Edition, 9876-9*
	Electronic Devices, 20157-8
	Essentials of Electronic Devices, 20062-8
	Principles of Electric Circuits, 2nd Edition, 20402-X
	Electric Circuits, Electron Flow Version, 20037-7
STANLEY, B. H.	*Experiments in Electric Circuits, 2nd Edition, 20403-8*
BERLIN	*Experiments in Electronic Devices, 20234-5*
GAONKAR	*Microprocessor Architecture, Programming, and Applications with the 8085/8080A, 20159-4*
LAMIT/LLOYD	*Drafting for Electronics, 20200-0*
LAMIT/WAHLER/HIGGINS	*Workbook in Drafting for Electronics, 20417-8*
NASHELSKY/BOYLESTAD	*BASIC Applied to Circuit Analysis, 20161-6*
ROSENBLATT/FRIEDMAN	*Direct and Alternating Current Machinery, 2nd Edition, 20160-8*
SCHWARTZ	*Survey of Electronics, 3rd Edition, 20162-4*
SEIDMAN/WAINTRAUB	*Electronics: Devices, Discrete and Integrated Circuits, 8494-6*
STANLEY, W. D.	*Operational Amplifiers with Linear Integrated Circuits, 20090-3*
TOCCI	*Fundamentals of Electronic Devices, 3rd Edition, 9887-4*
	Electronic Devices, Conventional Flow Version, 3rd Edition, 20063-6
	Fundamentals of Pulse and Digital Circuits, 3rd Edition, 20033-4
	Introduction to Electric Circuit Analysis, 2nd Edition, 20002-4
WARD	*Applied Digital Electronics, 9925-0*
YOUNG	*Electronic Communication Techniques, 20202-7*

Survey of Electronics

3rd Edition

Leland P. Schwartz
Rio Hondo College

Charles E. Merrill Publishing Company
A Bell & Howell Company
Columbus Toronto London Sydney

Published by
Charles E. Merrill Publishing Company
A Bell & Howell Company
Columbus, Ohio 43216

This book was set in Univers.
Production Coordination: Constantina Geldis
Cover Designer: Cathy Watterson

Library of Congress Catalog Card Number: 84-62098
International Standard Book Number: 0-675-20162-4
Printed in the United States of America
 2 3 4 5 6 7 8 9 10—89 88 87 86 85

Cover photo: Blue wafers. By Joan Kramer and Associates/Clark Dunbar

Preface

The original objective in the writing of this book was to present an overview of the field of electronics which

1. Was elementary but written for adults.
2. Avoided rigorous mathematics.
3. Explained the operation of common electrical and electronic systems before dwelling upon the details of components.
4. Emphasized the use of analogies as a means of gaining new concepts.
5. Included only essential material, thus saving the instructor the problem of customizing the text but encouraging augmentation and amplification from personal experience.

In this third edition, all of the above have been retained and appropriate optional experiments have been added. Up-to-date developments in the field of electronics are included and many sections of the book have been modified to improve clarity. As a result of this new material, some information about resistors is presented in an early section. In addition, extensive lists of fill-in-the-blank questions have been placed at the end of each chapter to aid students in determining if they have assimilated the material presented. Instructors will find that these questions can be used as homework assignments or quizzes and that the questions will require a minimum amount of time to grade.

Three groups of students will especially benefit from this book. First are the liberal arts majors who want a science course, with or without lab, to round off their education. In the second group, majors in various technologies, such as radiology, nursing, drafting, automotive, and computer or appliance repair, may need a brief background in electronics to function better in their own fields. The third group is the newest; these students may seriously be interested in pursuing electronics as a career, but they have had little or no prior experience or training in the subject and, therefore, lack the background necessary to survive a standard college-level electronics course in DC/AC circuits without some help.

In my experience, these three groups respond especially well to presenting systems first, then the details; therefore, I have organized this book in that manner. Students respond better to detail after they have seen a need for it. However, you may rearrange the chapter sequence if you'd prefer to use a more traditional detail-to-system approach.

Although applications to computers are mentioned, this book concentrates on more traditional electronics for two good reasons: (a) students, especially nonmajors, are more familiar with appliances such as radios and televisions; and (b) a radio lends itself better to a systems treatment in teaching basic, overall electronic (especially analog) principles.

I am grateful to the many people who have contributed ideas, time, and encouragement during the preparation of this manuscript. I give special thanks to Chris Conty for his persistance, patience, and perception; to Michael Chamberlain for his care and expertise in technical editing; and to Connie Geldis for promptly and courteously coordinating the production of the book.

I especially appreciate the help of Julio Ahumada of Monroe Community College, George Bramlett of the College of San Mateo, James Brittain of East Texas State University, Bert Evans of Diablo Valley College, Arch Gillespie of Maricopa Technical College, Bob Laurensen of Parkland College, Michael Pelletier of Northern Essex Community College, Mack Whitehurst of Pitt Community College, and Del Yaeger of East Texas State University. Most of all, I thank those students whose questions identified the sections of the earlier editions of the book that were most in need of revision.

Leland P. Schwartz
Whittier, California

Contents

6   *A RADIO TRANSMITTER*   **75**

Part II   *Components and Theory*   **123**

10   *POWER, RESISTANCE, AND RMS*   **125**

11   *METERS*   **149**

xii

Part

Devices

1

A Few Definitions

Electricity has been known for over three hundred years; the term *electronics* has been used for less than fifty. What, in recent years, has made study in this area so exciting? Furthermore, what is the difference between electronics and electricity?

In modern usage, *electricity* is generally the term applied to large amounts of electric power, such as that used for motors, lighting, and heating. Much excitement surrounds new ways to generate such power. For example, in the desert near Daggett, California, an experimental 10-megawatt power plant makes use of acres of mirrors (heliostats) to concentrate reflected sunlight onto a boiler atop a tower where steam is produced to power turbine generators (Figure1.1). In other locations, experimental new versions of windmills are being developed to drive such dynamos.

The word *electronics* is usually reserved for electrical devices that perform delicate control tasks. For many years most people as-

FIGURE 1.1   Ten-megawatt experimental power plant

sociated the word with little other than radio and television. Today, however, use of electronics has exploded into almost every area in our lives.

If you visit a doctor's office or a hospital you may be X-rayed, metered, or treated by electronic equipment. Your malady may be diagnosed with the aid of an electronic computer located in a distant city, with all communication done electronically. Some of this may be by conventional wire cables, but more and more new installations use laser signals carried by glass fibers. Typical electronic medical equipment is illustrated in Figure 1.2.

In your home you may have electronic games or a home microcomputer, plus several pocket calculators, digital clocks, a microprocessor-controlled oven or other appliances, and (perhaps) an electronic organ or guitar (Figure 1.3). In addition, you probably have the usual radio, television, audio or video tape recorder, record player, and telephone.

Already, many products are being machined, assembled, or packed by robots. In fact, this new and most exciting field, called **robotics,** promises to be an area of great growth in the years to come.

The common existence in most homes of so many electronic devices demonstrates the overwhelming influence electronics now

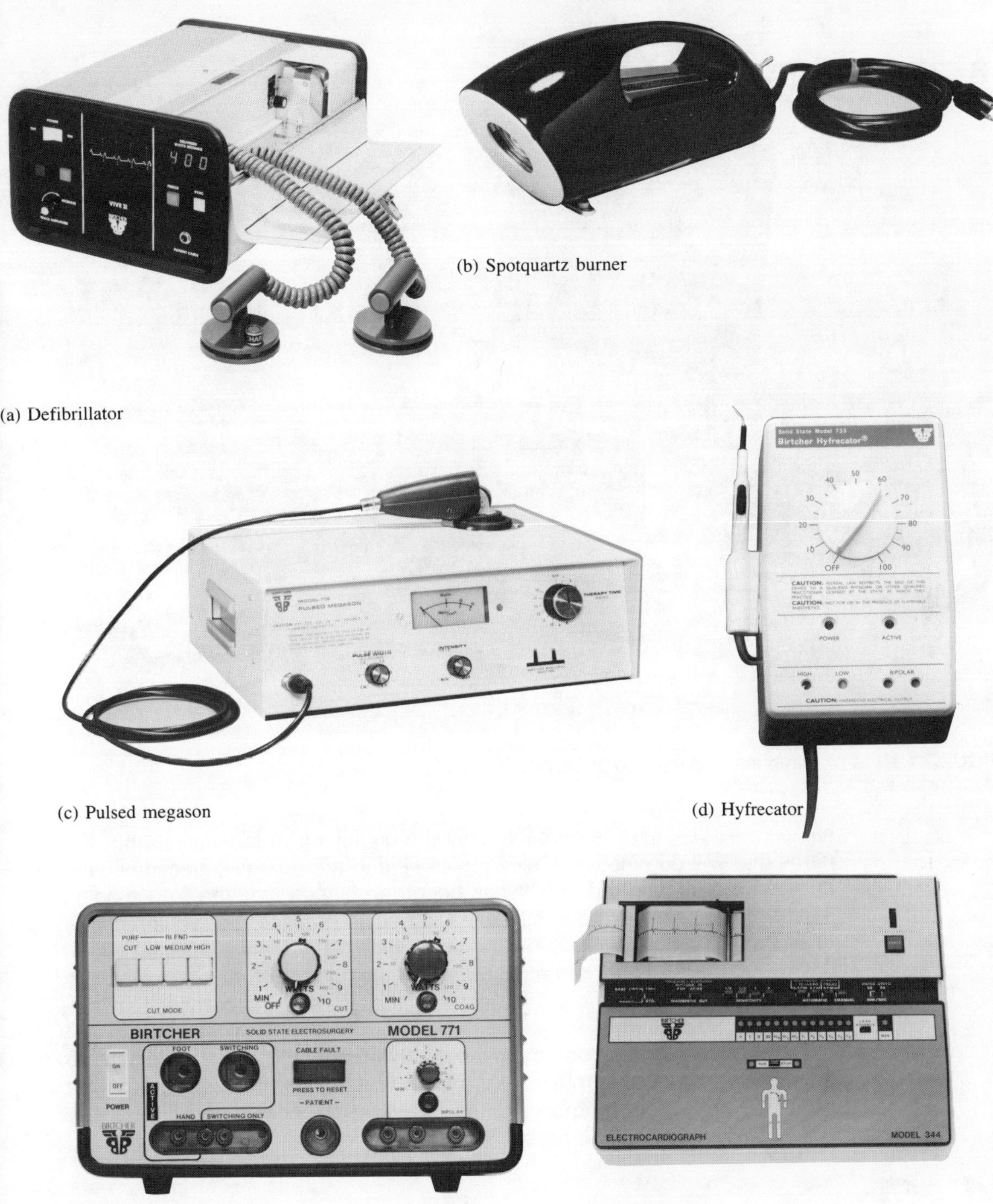

(a) Defibrillator

(b) Spotquartz burner

(c) Pulsed megason

(d) Hyfrecator

(e) Electrosurgery instrument

(f) Electrocardiograph

FIGURE 1.2 Typical medical electronic instruments *(Courtesty of Birtcher Corporation)*

FIGURE 1.3   Microcomputer *(Courtesy of Morrow Designs)*

has on our lives. This is probably only the beginning. It is possible that the items that will dominate the electronics field in the next two decades have not even been invented yet. It has become obvious that the future holds a substantial need for persons skilled in design, service, and maintenance of electronic products.

It is important to note that electricity and electronics share the same basic scientific principles. It is the element of **control** that distinguishes electronics.

In this book there will be examples used that could fall into either category. The emphasis will be on electronics, but one should not be dismayed when common electrical devices are considered.

1.2 *ELECTRONIC TERMS*

It may be necessary to learn many terms while proceeding through the text. To help you learn them, each is printed **boldface** the first time it is mentioned and defined.

SYSTEMS 1.3

Part One of this book is organized on a basis of progressively more complex systems. Each system will be represented by a **block diagram,** which will show how **subsystems** are assembled together to form a total system that will perform a desired function. For instance, a flashlight is an electrical system designed to provide a portable source of light. In block diagram form it would appear as in Figure 1.4. The battery, switch, and bulb are actually subsystems whose functions are generally of value only when combined with other subsystems.

TRANSDUCERS 1.4

Most systems contain one or more **transducers.** A transducer converts energy from one form to another. For example, a microphone converts sound (mechanical energy) into electromagnetic impulses (electrical energy), and a loudspeaker converts electrical impulses back into sound.

In the block diagram of Figure 1.4, the battery and bulb are both, technically, transducers. The battery converts chemical energy to electrical energy, and the bulb converts electrical energy to light.

The basic forms of energy are mechanical, electrical, light, heat, chemical, and nuclear. If a machine were made to use the energy in ocean waves to pump water, it would not be considered a transducer because the wave energy and the energy used in elevating the water are both forms of mechanical energy.

AMPLIFIER 1.5

Another subsystem frequently found in electronic systems is the **amplifier.** This is a device in which a small amount of energy input is made to

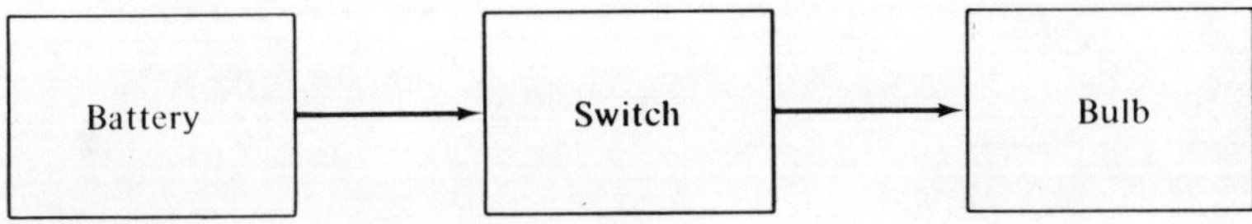

FIGURE 1.4 Flashlight block diagram

look like a larger amount in the output. A small impulse signal controls a large source of energy in such a manner that the signal is approximated in form in the output, but the magnitude (strength) is changed. For instance, if the original signal looked like the letter "m," the output would look more like "𝔪." Note that the new form is taller but not wider. This is the case because the "picture" of a signal is usually a graph of magnitude plotted against time, and while the strength of the peaks and dips are magnified in the output, these changes still occur in the same amount of time (Figure 1.5).

1.6 *ELECTRONS*

It is as unlikely that electronics could be discussed without **electrons** as to mention painting without considering paint. Electrons are extremely tiny but numerous "particles" that can go through seemingly dense substances, such as metals. They have been arbitrarily named *negative particles* to distinguish them from **protons,** which are positively charged, and **neutrons,** which have no electrical charge. It is convenient to identify electrons with the negative sign (−) and protons with the positive sign (+). There is no mathematical significance in these labels except that they indicate opposites. They could just as well have been called male ($\male$) and female ($\female$) to indicate the difference.

To understand how we can control electrons to do useful work requires some knowledge of atomic theory. Although atomic theory has been extensively modified, it was first hypothesized in 1913 by Dr. Neils Bohr, a Danish scientist. Basically, the idea is that the primary structure of matter is the atom, which consists of a **nucleus** made up of protons and, usually, neutrons, the whole of which is surrounded by electrons like planets around a sun. While this analogy is quite useful because it is easily visualized, it is important to note that modern physics recognizes

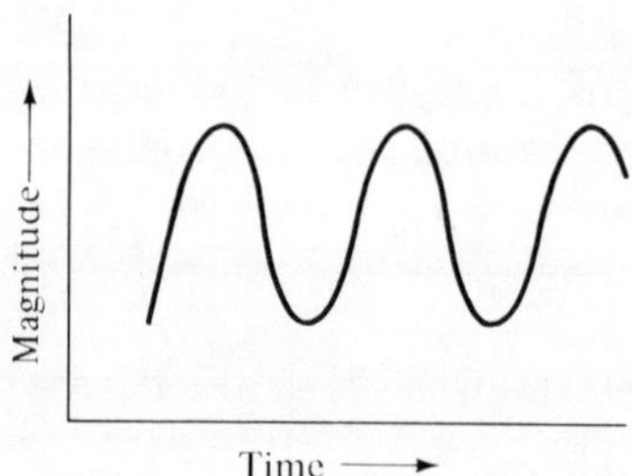

FIGURE 1.5   Signal graph

electrons exhibit a "wave nature" not unlike light or sound energy waves, as well as the conventional material aspect of charged particles that we discuss here.

Hydrogen is the most elementary atom. In its most common form, it consists of just one proton for a nucleus and has one electron that whirls around it.

Helium is slightly more complex. In addition to two neutrons, the nucleus contains two protons. Two electrons spin around the atom.

Perhaps you have already noticed that with these two elements, the number of electrons is equal to the number of protons in the nucleus. This fact is true of all atoms in their stable form. Whenever an atom has more or less electrons than it has protons, it is said to be an **ion**. In this state it is unstable, and therefore, very active.

Electrons orbit around their nuclei at very high speeds and would be projected off into space by centrifugal force if it was not for the attraction between the positive charges in the nucleus and the negative charges of the electrons. There is a very critical balance between these forces that allows the electrons to exist at certain discrete distances from the nucleus. These distances are called **shells** because they surround the nucleus like the shell of an egg surrounds the yolk.

The hydrogen and helium atoms have electrons only in the first shell from the nucleus. This is usually called the *K shell* and can accommodate no more than one pair of electrons. Other atoms have electrons which occur in shells farther from the nucleus, which are called *L, M, N,* etc. Each of these are subdivided into **subshells**. The second shell (*L*) has two subshells, the third shell (*M*) has three subshells, the fourth (*N*) has four subshells, and so on. In each case, the first subshell will accommodate no more than one pair of electrons. The second subshell will allow up to three pairs, the third allows five pairs, and so the pattern continues by odd numbers.

As was previously mentioned, electrons can move through apparently dense substances. Materials through which electrons drift with relative ease are called **conductors**. Those which present great opposition to passing electrons are called **insulators**. Good conductors, like copper (shown in Figure 1.6), have only one or two electrons in the outermost subshell. Elements that are poor conductors or insulators have that subshell nearly or completely filled. Elements with four electrons in the outermost subshell are neither good nor very poor conductors. They are, therefore, called **semiconductors**. Among this group of elements are silicon and germanium.

When the number of negatively charged particles at a location is equal to the number of those positively charged, the area is said to be of **neutral polarity**. If either kind of charge exceeds the other, the area is said to be **polarized**. That is, if the number of electrons present exceeds the number of protons, the polarization is negative. More positive charges than negative ones would exhibit positive polarity.

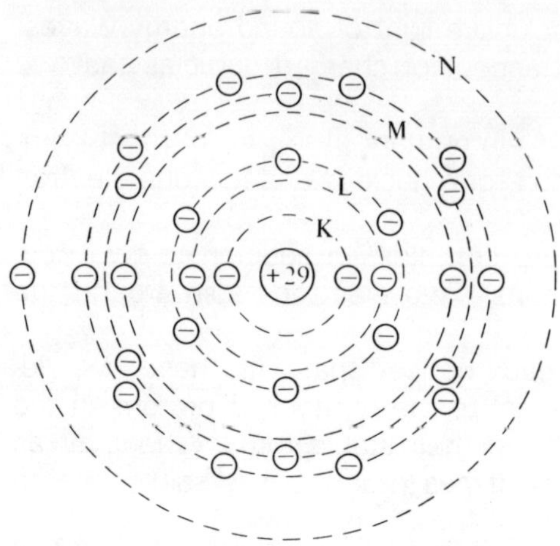

FIGURE 1.6   Diagram of copper atom

1.7 *MAGNETISM*

From time to time **magnetism** will be involved in the devices and functions that will be discussed. Its existence with electronics may seem
somewhat mysterious at first, but like electrons, it will be discussed more
in later chapters. At this point it is sufficient to know that it coexists with
electronics and is important to it but not the same thing. Magnets are
objects that always exist with two opposite **poles**. These could have been
named *male* and *female*, *yes* and *no*, *from* and *to*, or any other names
that would show how opposite they are, but it is customary to call one of
them the **north pole** and the other **south pole**. The north pole is usually
marked N, and the south pole S. This labeling resulted from the discovery
that when a magnetized bar is allowed to move freely, one end swings
toward the North Pole of the earth. This is the one called the *north pole*.
It would be more accurate to call it the *north-seeking pole*, but common
usage has shortened the expression.

1.8 *REPULSION AND ATTRACTION*

Electrons repel other electrons but are attracted to protons. Protons also
repel each other but are frequently locked together by other nuclear forces.
Two magnetic north poles will repel each other, as will two south poles;
however, north and south poles are strongly attracted to the other. The
above can be simply stated as follows: *Likes repel; unlikes attract.*

REPULSION AND ATTRACTION

Materials needed

1—Glass rod (1/4″ to 3/4″ diameter, 6 to 12 inch length)
1—Piece of silk or rayon cloth
1—Hard rubber rod (about same size as the glass rod)
1—Piece of cat fur
2—Bar magnets
1—Ounce iron filings
1—Pair pith balls connected by about 10 inches of thread
1—8 1/2″ × 11″ sheet of cardboard
Several bits of newsprint (1/8″ across or less)

Note substitutions:

Glass rod—Barkeeper's stirring rod
Hard rubber rod—Hard rubber or plastic comb
Cat fur—Dry head of human hair
Iron filings—A magnet will glean these from grinding wheel dust.
Pith balls—Styrofoam packing "peanuts"

PROCEDURE

1. Note that bar magnets cling together when the N pole of one is touched to the S pole of the other, but repel each other if N ends or S ends are pushed together.

2. Lay one magnet on a flat surface and place the sheet of cardboard over it. Sprinkle iron filings on the cardboard and gently tap so that the filings can shift positions. Note the pattern the filings make showing the presence of the magnetic field between the two ends of the magnet.

3. Place the two magnets so that the N pole of one is an inch or so from the S pole of the other. Again cover with cardboard and sprinkle iron filings to trace the magnetic field.

4. Reverse one of the magnets and repeat the previous step.

5. Briskly rub the hard rubber rod with the cat fur. While holding the center of the thread connecting the pith balls so that the balls hang together, touch the balls with the rod. Observe that they are at first attracted to the rod, then repelled by it and by each other. (The rod will be negatively charged.)

6. Rub the glass rod with the silk or rayon cloth. Repeat the previous step using the glass rod instead of the hard rubber. (The glass rod will be positively charged.)

7. Try to pick up pieces of paper with the rods. Note that at first a pickup may be made, but moments later the paper is ejected.

QUESTIONS

Select a word or words from the following list to fill each of the blanks in the questions.

amplitude	four	semiconductor(s)
amplifier	negative	seven
conductor(s)	neutral	six
control	nine	south
eight	north	subsystem(s)
electron(s)	one	strength
insulator	positive	system
ion	seeking	three
five	selection	transducer

1. An amplifier increases the ___________ of a signal.

2. Primarily, electronics differs from electricity only by the matter of emphasis on ___________.

3. A block diagram provides a picture of the relationship between ___________ in a ___________.

4. Conversion of a signal from one form of energy to another is accomplished by a ___________.

5. An electron has a ___________ charge.

6. A proton has a ___________ charge.

7. A neutron has a ___________ charge.

8. The south pole of a magnet is attracted to a magnetic ___________ pole.

9. A device that permits a small force to control a large source of energy is called a(an) ___________.

10. It would be more meaningful if a magnetic north pole was called a north ___________ pole.

   A FEW DEFINITIONS

11. A positive pole will attract __________ charges called __________.

12. The M shell of an atom contains __________ subshells.

13. Elements that contain only one or two electrons in their outermost subshells are good __________.

14. Semiconductor elements contain __________ electrons in their outermost subshells.

15. An atom in which the number of electrons is unequal to the number of protons is called an __________.

2

Elementary Units

In Chapter 1 (Section 1.3), a block diagram was illustrated for a common flashlight as repeated in Figure 2.1

This example provides a convenient basis for examining some of the elementary components necessary to nearly all systems. For instance, there must be some source of power to activate the device. With the flashlight, this source is a battery.

A battery is a device consisting of one or more **cells**. A cell is a basic unit (or module) that will produce an **electrical potential**. The proper name for

15

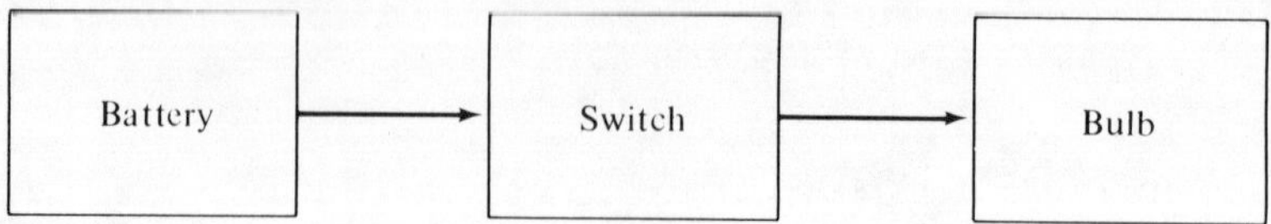

FIGURE 2.1 Flashlight

this energy is **electromotive force**, but since it is such an unwieldy name, it is frequently abbreviated to **EMF**. Because it is measured by a unit called the **volt**, the term **voltage** is commonly used.

Electrical potential or **potential difference** occurs when electrons are more numerous at one terminal of the cell than at the other. The **terminal** (contact) with this greater number of electrons is called the **cathode** and has a **negative polarity**. The other terminal, the **anode**, has **positive polarity**. The greater the existing differences are in the proportion of negative charges to positive charges, the greater is the potential (force) that exists between the terminals. This is true because the electrons strain to get away from each other more as they are crowded closer together at the negative terminal. Furthermore, the fewer the number of electrons that are at the positive terminal, the greater is the attraction for the electrons at the negative terminal. It is convenient to consider this "straining" of electrons to flow from negative to positive terminals to be like an electrical difference in pressure. The difference in potential may also be referred to as **voltage drop**. This term may seem confusing at first (particularly when used to indicate a change from a negative voltage to a more positive one). However, such usage will be clarified in later chapters.

Battery cells are of many types, but probably the most common are **chemical cells**. The chemistry of their operation is quite interesting, but beyond the scope of this text. Basically, chemical cells produce an EMF at their terminals because of a chemical reaction between a fluid called an **electrolyte** and the surfaces of two dissimilar materials. One of these surfaces acquires a surplus of electrons due to the chemical action and thereby becomes very negative, whereas the other surface suffers a deficiency of electrons and becomes very positive.

The chemical action of the cell is like that of a pump. This can best be explained by analogy. For instance, suppose two tanks of water are connected by an underground pipe as shown in Figure 2.2. Gravitational pull and free flow cause the water level in each to be the same. If a pump is now installed and operated in the connecting pipe, the water level in one tank will rise and the other will fall, as illustrated in Figure 2.3. The difference in their levels is a potential difference that would permit the water from the higher level to flow to the other if the pipe were reconnected (without the pump) between them.

With the battery cell, the same principle applies. If provided with a wire "pipe" through which they can travel, the electrons on

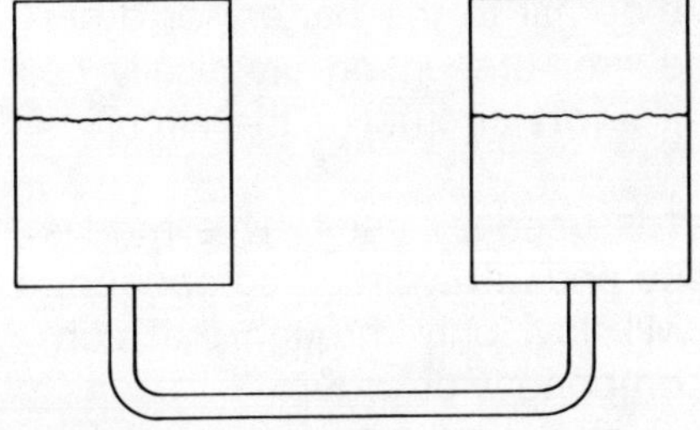

FIGURE 2.2   Water tanks

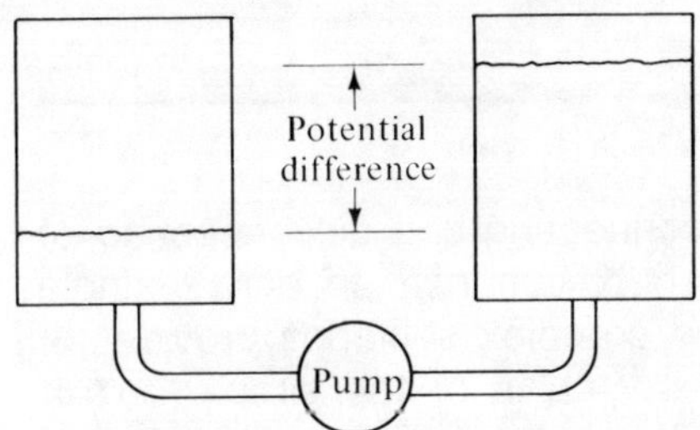

FIGURE 2.3   Difference in water levels

the cathode will flow back to the anode because of the potential difference. With electricity, this is called *closing the circuit.*

ELECTRICAL CIRCUIT **2.3**

Continuous electron flow can take place only in a **closed circuit**. That is,
there must be a continuous loop so that the electrons can flow out of the
electron pump (battery, generator, or other device), through conductors of
some type, and back to the pump's input. No electrons are lost in the
process. They simply circulate around and around through the circuit. Usually they are forced to do some kind of work between the output and input.
A somewhat similar closed circuit system is the cooling system of the
automobile. The water circulates (driven by a pump) through the engine
where it picks up heat. It is then conducted through the radiator where
the heat is released and the water returns to the pump.

If at any time an electrical circuit is broken so that a complete loop is no longer formed, the condition is called an **open circuit**.

Occasionally a portion or all of the circuit external to the battery or other power source may be bypassed by a conductor presenting practically no opposition to electron flow. This is called a **short circuit**, or the circuit is said to be **shorted**.

Sometimes the word *short* is used by itself. This means that a conductor connects points on opposite ends of a circuit component, or group of components, so that current will flow only through that conductor (or short) rather than through the component or group.

2.4 *SWITCHES*

A switch is any device that permits the connecting and disconnecting of a path for electron flow. The simplest variety connects to only a single path and is called an *on/off switch*. More complex switches provide for connecting multiple paths. The flashlight switch is of the simple on/off type that can connect a conductor from the battery to the bulb. It is a **single-pole, single-throw (SPST)** type. That is, there is only one terminal (pole) to be connected or disconnected with the battery terminal. A connecting link swings only to or from that one pole, so it is termed *single throw*. The symbol is illustrated in Figure 2.4. The symbols of a few other common switches are also shown.

2.5 *INCANDESCENT BULBS*

The flashlight bulb is a glass enclosure containing a very thin conductor, called a **filament,** that has each end connected to a terminal **insulated** from the other on the bulb's base. As mentioned in Section 1.6, insulation is a material that will not readily permit electrons to flow through it. The inside of the bulb is evacuated of air and may be filled, instead, with a gas that cannot chemically react with the filament. When electrons flow through the filament in large quantities, they encounter a form of **impedance*** called **resistance**. This is something like the experience you may have had when a rope was pulled through your hands as you gripped it. As you painfully

---

*There are two other forms of impedance that will be explained later.

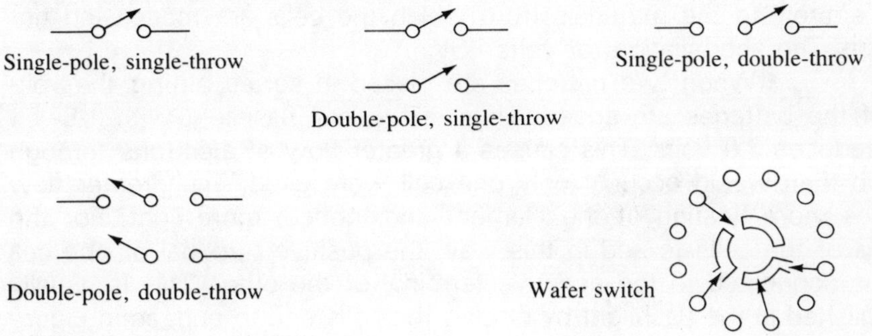

FIGURE 2.4 Switch symbols

found out, the friction created produced heat and you suffered a rope burn. Similarly, as electrons course through the filament, the friction-like resistance of the filament material causes it to become very hot. Just as heating any material to a high temperature causes it to glow, the filament glows (or becomes **incandescent**).

One of the terminals of the bulb is connected to the negative terminal of the battery (the bottom metal plate or base) through the switch. The other bulb terminal is connected to the positive terminal (the small metal cap in the center top of the cell). Thus a closed circuit is produced as illustrated in Figure 2.5.

VOLTAGE **2.6**

As was mentioned in Section 2.2, battery cells produce an electromotive force (EMF) measured in volts. Ordinary dry cells such as those used in flashlights produce 1.5 volts, whether they are large or small. The voltage

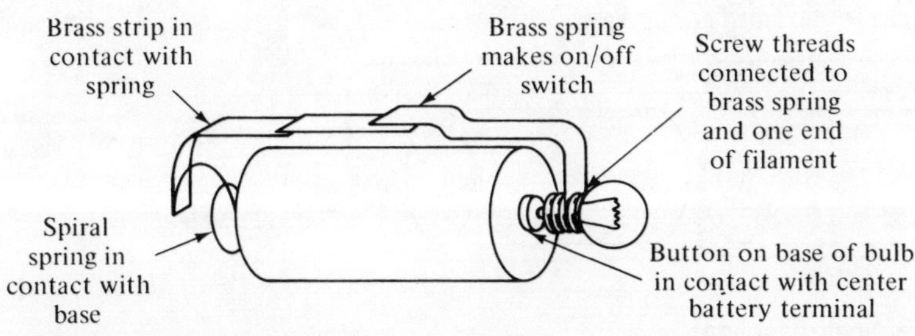

FIGURE 2.5 Flashlight bulb and battery

is due entirely to the materials from which the cells are made, and not their size. The abbreviation for volts is *V.*

When two batteries are placed in **series aiding** the voltages of the batteries are added. Thus, a two-cell flashlight with 1.5-volt cells produces 3.0 volts. This causes a greater flow of electrons through the bulb than would occur if only one cell were used. The greater flow produces more heating of the filament and, hence, more light. For the voltages of the cells to add in this way, the positive terminal of one cell must be connected to the negative terminal of the other. This is usually accomplished in the flashlight by placing the cells end to end, as in Figure 2.6, so that the case (negative terminal) of one cell contacts the center terminal (positive) of the other. As more cells are stacked on one another in this way, the voltage is increased proportionately. Thus, three cells produce 4.5 volts, four cells make 6.0 volts, and so on.

Of course, batteries and other voltage sources may also be placed in **series opposing**. A typical situation illustrating this is the circuit for charging a battery. The like terminals of the charger and battery are connected so that the greater charger voltage "bucks" the flow of the lower battery voltage causing electrons to flow *into* the negative terminal and *out* of the positive terminal of the battery. The total voltage across the combination will be the difference of their respective voltages. Suppose, for instance, that a 6-volt battery is in series opposing with one of 2 volts. The total voltage across the combination would be 4 volts.

Since a small dry cell provides the same voltage as a large one, a person might be led to think that size does not matter. However, the power a cell can deliver depends not only on voltage but also upon the number of electrons per second it can provide. If the rate of drain of electrons is too great for a cell's capacity it will be destroyed. Where the required power drain is great, large cells are used (see Figure 2.7), or cells are put in **parallel**.

When cells are placed in parallel, two or more are connected to a load so that their positive terminals are **common** (connected together) and their negative terminals are common.

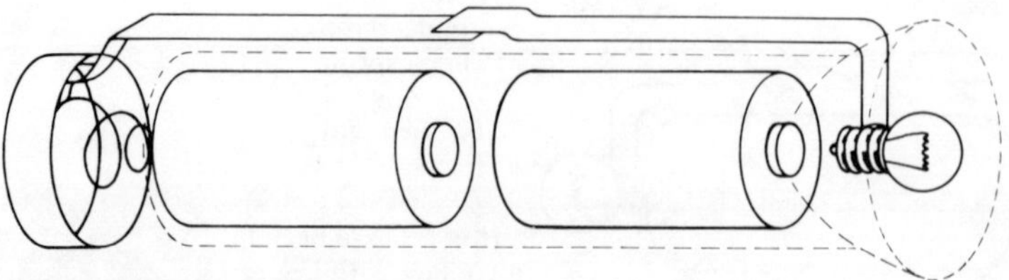

FIGURE 2.6 Two-cell flashlight

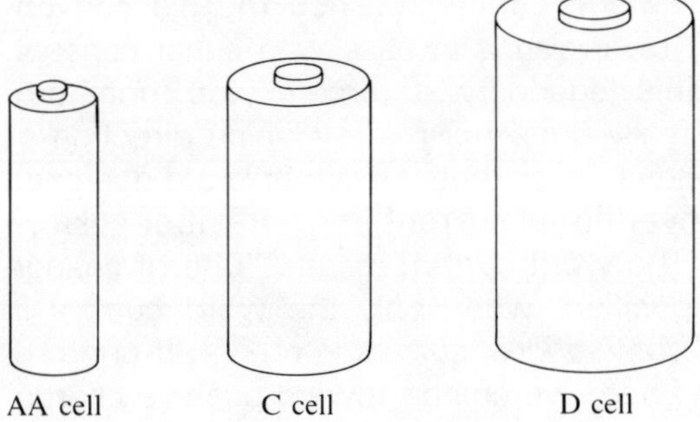

FIGURE 2.7 Dry cell size comparison

CURRENT 2.7

When voltage in a closed circuit is increased, the flow of electrons is increased. That is, more electrons are caused to move past any given point in the circuit each second. They individually do not move any faster than before, but there are more of them going by each second. This is called an increase in *rate of change of charge.* More commonly, it is called an increase in **current**. Current can be likened to sand being carried on a conveyor belt. No matter how much is loaded on the belt, it travels at the same speed. The amount carried past any point per second depends on how heavily the belt is loaded. Similarly, an increase in the number of electrons that flow past a point in a second means that current has increased.

Over two hundred years ago Benjamin Franklin recognized that some kind of flow took place between oppositely charged terminals. In fact, it was he who named one of the terminals *positive* and the other one *negative.* It appears that positive was applied to the charge of an object from which arcing or ''electrical fire'' was observed when two different materials were rubbed together. Franklin believed that this electrical fire was ''an element diffused among, and attracted by, other matter, particularly by water and metals.'' However, he was aware that he had no proof of the direction of movement. Legend states that Franklin's response to questions about the assumed direction of flow was that it does not matter what direction is assumed provided that you are consistent.

The assumption of flow from positive to negative seemed to please everyone for over one hundred years. Then, around 1900, it was discovered that electrons were negatively charged particles and, further, that it was electrons that were moving in a conductor. As you know, like charges repel, so that meant that electrons were flowing in the direction opposite to Franklin's assumed direction of current. Many people immediately changed their concepts of flow in order to think and express themselves in terms of electron movement. Others retained the old conventional current direction. The smart technician is at ease with either concept. It now appears that Ben Franklin's legendary statement was more profound than originally realized. It really does *not* matter which direction of flow is assumed, provided that one is consistent! All symbols are designed on the basis of what is called **conventional current** (movement of positive charge toward negative charge). The word *current* means "rate of change of charge." Unless otherwise specified, when only the word *current* is used, it should be assumed that *conventional current* is meant. If one prefers to speak of the movement of negative charge toward positive charge, it is quite proper to speak of **electron flow**.

A single charge is extremely small, so most of the time large quantities are handled. Just as wheat is usually handled by the bushel rather than by the individual grain, charge is handled by the **coulomb**. One coulomb is a huge number* of charges. Current is the rate at which charge is moved. The unit of measurement is the **ampere**, which is defined as a current of one coulomb per second. The abbreviation for amperes is *A*.

2.8 *DIRECT CURRENT AND ALTERNATING CURRENT*

When electrons move in only one direction in a circuit the flow is termed **direct current (DC)**. This is typical of current obtained from a battery. The voltage from such a source is constantly of the same polarity and is called a **direct current voltage**. In fact, the invention of the chemical cell about A.D. 1800 by Alessandro Volta made low-voltage direct current the earliest practical use of electricity.

About a hundred years ago Nikola Tesla and a few others developed the use of **alternating current (AC)**. Current is said to be alternating if the electron flow is first in one direction and then, shortly thereafter, in the opposite direction. The electricity delivered by the power company is usually AC. For more detail see Section 5.5. Current alternates only because the applied voltage polarity is continually changing. Therefore, a source that would cause current to alternate is called **alternating current voltage**.

---

*One coulomb = 6.25×10^{18} charges.

IMPEDANCE AND RESISTANCE   **2.9**

In discussing what made the flashlight bulb glow, it was mentioned that the filament had a form of impedance called resistance. This term was explained as a sort of friction present in the filament that impeded (gave opposition to) the flow of electrons. Since different materials have different impedances, a unit of measure has been created for this characteristic. The unit is called the **ohm**. One ohm has been established as that amount of impedance that will permit one ampere of current when subjected to the EMF of one volt. The form of impedance that is normally encountered with a battery source of voltage is resistance. As with any form of impedance, resistance is measured in ohms. The symbol for ohms is the Greek letter omega, Ω.

Experiment **2**

VOLTAGES IN SERIES AND PARALLEL

Equipment needed

1—DC power supply
2—D dry cells and holders
1—Current meter (500-milliampere)
1—Voltmeter
4—10-Ohm, 1/2-watt resistors
1—Breadboard (surface on which circuit can be assembled)

Safety note

The DC power supply used in this and subsequent experiments derives its power from an AC line. Such a source is always dangerous if lead-in cords are worn or frayed. If such a condition exists, call it to your instructor's attention before use.

PROCEDURE

1. Measure the voltage of each cell.*
2. Connect the power supply and the two cells in series aiding as illustrated in Figure A, and set the power supply output to 10 volts.

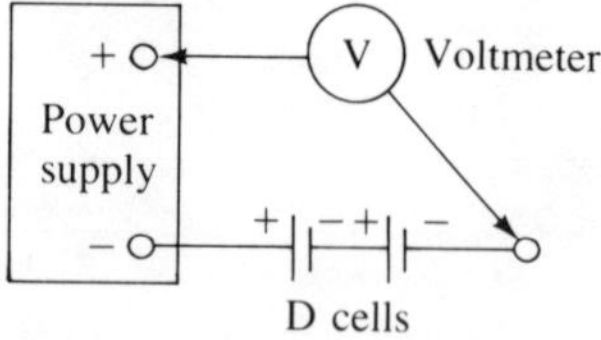

FIGURE A Cells in series

---

*When voltages are unknown, always start with the meter on a high scale, then work down to the lowest scale that will not "peg" the needle (drive it past the right end of the scale.)

  ELEMENTARY UNITS

3. Measure the combined voltage across the series circuit.

4. Reverse the connections of one of the cells.

5. Measure the combined voltage across the series circuit.

6. Connect the negative terminal of the current meter and one terminal each of the four resistors to a single point as in Figure B.

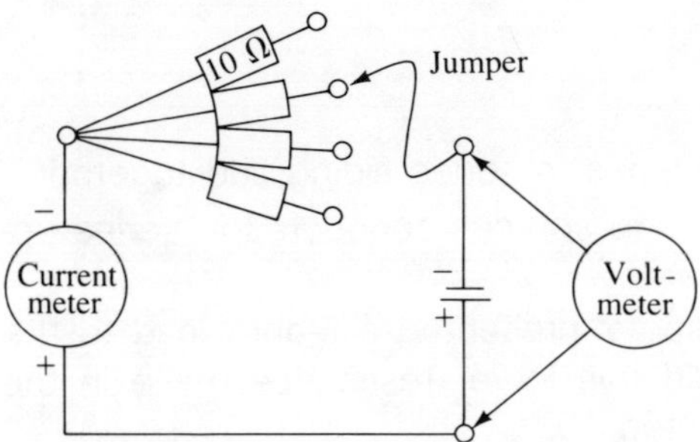

FIGURE B Loaded cell measurement

7. Connect the positive terminal of one cell to the positive terminal of the current meter.

8. Place the voltmeter across (in parallel with) the cell.

9. Momentarily contact a jumper (connecting wire) from the negative terminal of the cell to the open terminal of one 10-ohm resistor (Figure B). Do not keep this connection for more than 10 seconds. (Cell damage may result.)

10. Read the circuit current and the voltage across the cell.

11. Connect a jumper across the open terminals of two of the 10-ohm resistors.

12. Momentarily contact a jumper from the negative terminal of the cell to the paralleled resistors (Figure C).

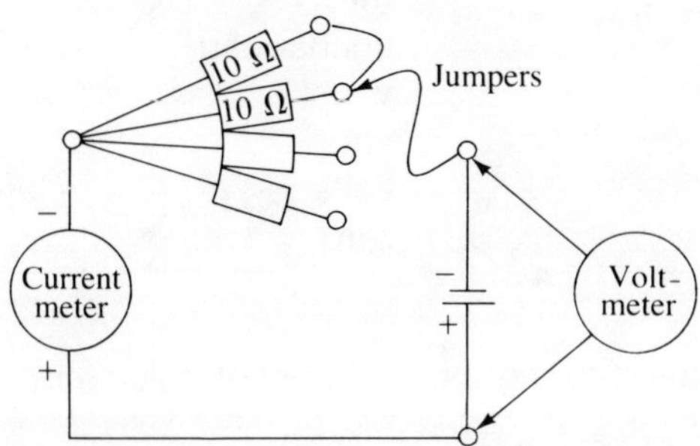

FIGURE C Greater cell loading

13. Read the circuit current and the voltage across the cell.

14. Connect all four resistors in parallel.

15. Repeat steps 12 and 13.

16. Connect the second cell in parallel with the first.

17. Repeat steps 9 through 15.

OBSERVATIONS

A. How did step 3 indicate that voltages in series aiding add together?

B. How did step 5 indicate that voltage sources connected in series opposing subtract from one another?

C. The two parallel resistors of step 12 presented a 5-ohm load to the voltage source. What change did this lower resistance make in the voltage reading compared with step 10?

D. Compare any differences in voltage readings for each load as observed for two cells versus one cell. Note that when the load resistance was lowered, the additional cell was more important.

QUESTIONS

Select a letter, symbol, word, or words from the following list to fill each of the blanks in the questions.

A	electrolyte	second
alternating	electrons	series
ampere	filament	short
chemical	increased	single
closed	insulator	sum
common	negative	throw
current	ohm	unchanged
decreased	Ω	V
difference	open	volt
direct	parallel	voltage
double	pole	zero
electrical	positive	

1. The batteries used in automobiles, flashlights, and pocket calculators are __________ cells.

2. A potential difference occurs between the terminals of a cell because the chemical action causes __________ to be removed from one terminal and crowded onto another.

3. Electromotive force is a kind of __________ pressure.

   ELEMENTARY UNITS

4. The _________ is the unit of measure of EMF.

5. Electrical potential is always a measure of _________ between two points, just as height is.

6. A dry cell consists of a cathode, an anode, and an _________.

7. The cathode is the _________ terminal and the anode is the _________ terminal.

8. Electrons flow because of a difference in potential between two points in a closed circuit. The potential difference is called EMF or voltage. The single word, _________, indicates that there is movement of charge.

9. Potential difference is sometimes referred to as "_________ drop."

10. An ampere is the movement of one coulomb of charge per _________.

11. The opposition to the movement of electrons is measured by the unit called the _________.

12. If two sources of EMF are in series aiding, the effect is the same as one source of EMF whose value is the _________ of the two.

13. If two sources of EMF are in series opposing, the voltage across the pair is the _________ of the two.

14. The symbol for the unit of impedance is _________.

15. When the terminals of a source of EMF are touched together, or when a conductor is placed in parallel with an impedance, the connection is called a _________ circuit.

16. Continuous current can flow only in a _________ circuit.

17. Cells may be placed in _________ to prevent shortened life due to excessive current demand.

18. When electrons are caused to move in only one direction, the flow is called _________ current.

19. When the polarity of a power source continually changes, the output is called _________ _________ voltage.

20. When referring to a switch, the letters SPDT stand for _________ _________ _________ _________.

21. The tiny conductor that glows inside an incandescent light bulb is called the _________.

22. A D drycell can provide more _________ than a C cell.

23. When two or more things are connected to the same conductive surface, their connection is said to be _________.

24. When current is increased, electrons move at _________ speed.

25. Electron flow is a movement of _________ charges toward a _________ pole, but conventional current implies a movement of charges toward a _________ pole.

3

Relationship of Voltage, Current, and Impedance

CURRENT VERSUS VOLTAGE 3.1

In the previous discussion of the simple system of a flashlight, we found that an increase in voltage achieved by placing batteries in series would cause an increase in current through the bulb filament. The more current that flowed, the brighter the lamp glowed. In fact, if the impedance of any device is held constant, the current will increase or decrease in direct proportion to the EMF applied. For instance, if an EMF of 5 volts produced 2 amperes of current in a device of constant resistance, 15 volts would produce a current of 6 amperes. Or, if the EMF were reduced to 1 volt, the current would be reduced to 0.4 ampere.

CURRENT VERSUS IMPEDANCE 3.2

It is also true that if the voltage of a circuit is held constant, current will be *inversely* proportionate to impedance. That is to say, when impedance is increased, current will decrease. If there is a decrease in impedance, current will increase. As an example, if an impedance of 4 ohms allows a current of 9 amperes when an EMF of 36 volts is applied, lowering the impedance to 2 ohms would allow a current of 18 amperes. Raising the impedance to 12 ohms would reduce the flow to 3 amperes.

Impedance is a general term that applies to three kinds of obstacles to current. In their pure forms they are called resistance, **inductive reactance,** and **capacitive reactance.** Each or any combination of the three can be called impedance. This is analogous to applying the general term *dog* to a collie, fox terrier, poodle, or any combination thereof.

Of the three forms of impedance, resistance is the easiest to understand. Furthermore, it is the only form that exists in a circuit where the current is unchanging. For these reasons, it is the only form that will be thoroughly investigated in the next eight chapters.

The control of current by variation of resistance is extremely important in many devices. For instance, the speed control of a sewing machine or golf cart, or the volume control on your radio or TV are merely variable resistances that control the amount of current.

3.3   *OHM'S LAW*

Georg Ohm, a German experimenter, discovered the relationships of current, voltage, and impedance that have just been explained. As a result, he created a mathematical model, or formula, that expresses this relationship. Quite justly, it is now called **Ohm's law,** and its general form is

$$I = \frac{E}{Z}$$

The letter I originally stood for the word *intensity* from the expression *current intensity.* Today this is referred to merely as *current.* Therefore, I stands for the number of amperes of current.

The letter E is used to denote EMF. It will frequently be referred to as *voltage* and interpreted to mean the number of volts applied to the circuit being discussed.

Z is used to represent impedance as measured in ohms. If the impedance is purely **resistive,** an R is substituted for the Z so that the formula becomes

$$I = \frac{E}{R}$$

Formulas use plus or minus signs to indicate addition or subtraction, but multiplication is indicated without an × just by placing letters together. Thus, AB means quantity A times quantity B. Division is shown like a fraction instead of by use of ÷. Therefore, E/R means E divided by R.

Ohm's law is the most fundamental relationship in electricity and electronics. The student should make a particularly conscientious effort to understand it before proceeding further into the study of this text.

FORMULA MANIPULATION RULES # 3.4

Formulas such as $I = E/R$ can be easily converted to other forms if one understands a few simple rules.* Since the formula is stating that I is equal to the quantity, E/R, the same amount can be added to I and to E/R and the equality will still be true. Likewise, an amount can be subtracted from both sides of the equal sign and equality will still exist. The same idea makes it possible to multiply or divide each side by the same (nonzero) number without destroying the equality.

Suppose we know I and R and wish to find E. We can look at the formula $I = E/R$ and observe that our unknown E is shown divided by R. E can be isolated by performing the opposite operation. That is, if both sides of the equal sign are *multiplied* by R, the result is $IR = E$. Therefore, $E = IR$. If the known values of I and R are now entered into the new formula, the value of E can be calculated.

Similarly, we might have a problem where E and I are known and for which we want the value of R. From the original formula, $I = E/R$, it can be seen that R is in the denominator of the right-hand term. We need to have it isolated and in the numerator. It is necessary, therefore, to perform an operation opposite to the division indicated. So, just as was done in the previous paragraph, both sides are multiplied by R. Again, the result shows $IR = E$, but we are not finished, because even though R is now in the numerator, it is still not isolated. Since multiplication by I is indicated, we must do an opposite operation. After *dividing* both sides of the equation by I the result is $R = E/I$.

Likewise, other formulas can be manipulated to obtain any unknown quantity within them expressed in terms of the others.

---

*Individuals skilled in algebra may omit this section.

3.5 *RESISTORS IN SERIES*

A very simple circuit such as the flashlight discussed in the previous chapter may consist of only one resistive element such as the bulb. However, it is very common to find circuits that employ resistors in series. If you ever had the experience of replacing a burned-out Christmas tree bulb, where the entire string went dark with the failure of one bulb, you worked with a series circuit. Other household devices using resistors in this arrangement are toasters and electric heaters. It is likely that the toaster in your home has four heating elements (one for each side of two slices of bread) that are in series. Many electric heaters have a fan placed in series with the heating element so that hot air is circulated when the heater is operating. Most electronic items include a number of resistances in series. Such frequent use of this circuit type makes it important to know how to calculate its characteristics.

A **series circuit** simply is one in which resistances are connected end to end so that all current flowing through one resistance also flows through the others. Figure 3.1(a) illustrates a series circuit composed of a battery, a fixed resistor, a **rheostat** (or variable resistor), and a lamp. Figure 3.1(b) is a schematic of the same circuit.

As shown in the circuit diagrams, all the electrons must go through each component. The total resistance of the circuit is easily obtained by adding the resistances. The rule for resistors in series is expressed by the general formula

$$R_T = R_1 + R_2 + R_3 + \ldots R_n$$

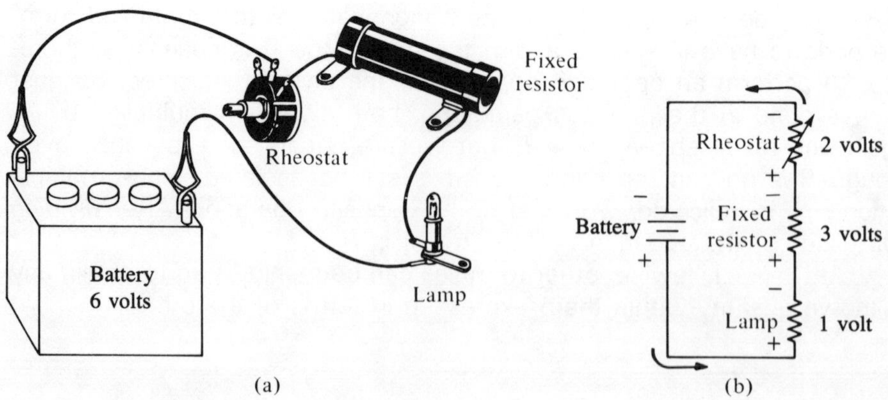

FIGURE 3.1   Resistors in series circuit

For example, for a fixed resistor of 12 ohms, the rheostat set at 18 ohms, and the lamp at 6 ohms, the total would be 36 ohms. With this information and a known applied voltage, the current in the circuit can be calculated by Ohm's law. Therefore, for a battery voltage of 6 volts,

$$I = \frac{E}{R} = \frac{6}{36} = \frac{1}{6} \text{ ampere}$$

Ohm's law also states that $E = IR$. Consequently, we calculate the voltage difference across each of the resistors by multiplying the circuit current by the resistance of each resistor. The voltage difference (voltage drop) across the fixed resistor would be

$$\frac{1}{6} \times 12 = 2 \text{ volts}$$

Likewise, the voltage drop across the rheostat would be

$$\frac{1}{6} \times 18 = 3 \text{ volts}$$

and the voltage drop across the lamp would be

$$\frac{1}{6} \times 6 = 1 \text{ volt}$$

The total would be 6 volts, the same as the supply voltage. It is important that in a series circuit, the summation of all the voltage drops equals the supply voltage. This idea is so fundamental that it has been given the special name of *Kirchhoff's voltage law,* which states: **The algebraic summation of voltage drops in a closed loop equals zero.*** To understand this statement, one must recognize that the polarity of an applied voltage is opposite to the polarity across each resistive component. To see this, please look closely at Figure 3.1(b). Electrons from the negative terminal of the battery first enter the rheostat and flow through toward the fixed resistor. Electrons move from a negative terminal toward a more positive terminal, so their entry point on the rheostat is marked to show a negative polarity and the exit terminal is marked to indicate positive polarity. Similarly the entry terminals of the fixed resistor and the lamp are shown to have negative polarities with respect to their exit terminals. From the negative to positive terminals of the rheostat is a +2 volts voltage drop; from negative to positive across the fixed resistor is another +3 volts; and from the negative to positive terminals of the lamp is still another +1 volt, for a total of +6 volts. However, as the flow is followed through the battery,

---

*Gustav Kirchhoff (pronounced "Kirk-hoff") was a great German physicist of the mid-nineteenth century.

the potential changes from positive to negative, or −6 volts. Now, as the voltage law indicates, the total is zero volts.

3.6 *RESISTORS IN PARALLEL*

Another common arrangement of resistors is called a **parallel circuit.** In your home, for instance, the lights and wall sockets are independent of one another although they are all connected to the main power line. Because a parallel circuit is electrically equivalent to the adjacent resistor symbols in a schematic diagram like Figure 3.2, they are said to be in parallel.

Electrons that leave one terminal of the power supply (Figure 3.2) may travel through any one of the resistors on their way to the other terminal. The resistance to their flow is, therefore, far less than if they each had to flow through all three, as in the series circuit.

Suppose, for example, that the resistance of the coffee maker was 40 ohms, the iron was 15 ohms, and the lamp was 120 ohms. Each would have 120 volts applied to it, and the individual currents would be calculated by the formula $I = E/R$, indicated thus:

$$\text{Coffee maker current} = \frac{120}{40} = 3.0 \text{ A}$$

$$\text{Iron current} = \frac{120}{15} = 8.0 \text{ A}$$

$$\text{Lamp current} = \frac{120}{120} = 1.0 \text{ A}$$

The total current supplied from the switch box would be the sum of the above currents, or 12 amperes. The total circuit resistance could be calculated from the formula, $R = E/I$:

$$R = \frac{120}{12} = 10 \text{ ohms}$$

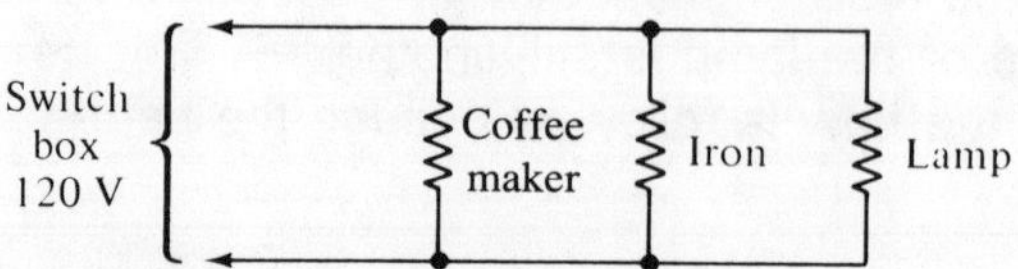

FIGURE 3.2   Resistors in parallel circuit

In this example the total resistance was very easy to calculate as soon as the total current was known. Unfortunately, the applied voltage or total current may not be known in all cases where knowledge of total resistance is desired. Therefore, an alternative method is needed. A formula for doing this exists, but it requires the use of **reciprocals.*** If your calculator has a reciprocal key (*1/X*), you are in luck. However, if it does not, don't despair. With many calculators a reciprocal can be found by entering the number, depressing the division key (÷), and then the equal (=) key. (If the calculator reads *1*, depress the equal key again.)

Total resistance can be calculated from the following formula:

$$R_T = \cfrac{1}{\cfrac{1}{R_1} + \cfrac{1}{R_2} + \cfrac{1}{R_3} + \ldots + \cfrac{1}{R_n}}$$

$$R_T = \cfrac{1}{\cfrac{1}{40} + \cfrac{1}{15} + \cfrac{1}{120}} = \frac{1}{.0250 + .0667 + .0083} = \frac{1}{.1000} = 10 \text{ ohms}$$

With most pocket calculators this is a very simple calculation to make. If you have the reciprocal key on yours use the steps in the following paragraph A. If you must use the method previously explained and have *memory plus (M+)*, *memory minus (M−)*, and *memory return (MR)* keys, follow the steps in paragraph B.

A. Enter the first resistor (40) and press the *1/X* key. Press the + key, enter the second resistor (15) and press the *1/X* key. Press the + key, enter the third resistor (120) and press the *1/X* key. Press the = key and follow this with the *1/X* key.

B. Enter the first resistor (40) and press the ÷ key. Follow this with the = key. (Press the = key again if a 1 shows on the register.) Press the *M+* key. Enter the second resistor (15) and press the ÷ key. Follow with the = key. Press the *M+* key. Enter the third resistor (120) and press the ÷ key followed by the = key. Press the *M+* key. All the reciprocals are now in the memory so press the *MR* key to retrieve them. Finally, press the ÷ key followed by the = key.

It is also possible to calculate a pair of resistors in parallel by the use of what is commonly called the *product over the sum* formula,

$$R_T = \frac{R_1 R_2}{R_1 + R_2}$$

---

*A reciprocal is 1 divided by the number. Thus, the reciprocal of 3 is 1/3, or of 5 is 1/5, etc.

In the previous problem the equivalent resistance of two of the resistances could be calculated as follows:

$$\frac{(40)(15)}{40 + 15} = \frac{600}{55} = 10.91 \text{ ohms}$$

This calculation could then be followed by another using the 10.91 ohms as one resistor in parallel with the 120-ohm resistance.

$$\frac{(10.91)(120)}{10.91 + 120} = \frac{1309}{130.9} = 10 \text{ ohms}$$

3.7   *RESISTOR CODING*

Fixed resistors are commercially available in many forms. The most common type is made of carbon. These have the advantages of being inexpensive, having fairly constant resistance over a wide range of temperatures, displaying low inductive and capacitive reactance, and being small-sized even for high resistance values. In addition, it is not difficult to vary the mixture of carbon and other materials to produce different values of resistance in the package. However, carbon resistors cannot be as accurately controlled to a specific value of resistance as can other types. When a value must be obtained to a **tolerance** of less than ±5% (plus or minus five percent) other types are almost invariably used.

Tolerance is a term very frequently used in most of the electrical and mechanical industries. It is impractical, if not impossible, to make components to exact measurement. The higher the precision required, the greater is the cost. Most applications do not require extreme precision, but some do. It is common practice, therefore, to specify the limits over and under the exact value that the component is intended to be. For instance, a resistor specified as 47 kilohms with a tolerance of ±10% should have a resistance of at least 47000 − 4700 = 42300 ohms but not more than 47000 + 4700 = 51700 ohms.

Modern carbon resistors are color coded to indicate their resistance value and tolerance. The color code is a convenient method of identifying resistors since the values can be read in any location. Of course, there are some disadvantages. Color-blind persons cannot read them, and resistors that have been overheated due to some malfunction may be so charred that the value cannot be determined.

The code consists of a series of bands, the first two of which indicate the first two digits of the resistance value. The third band indicates the placement of the decimal point or the number of zeros to

TABLE **3.1** Resistor color code

0	Black	
1	Brown	Gold = ±5%
2	Red	4th band only (tolerance)
3	Orange	Silver = ±10%
4	Yellow	
5	Green	
6	Blue	
7	Violet	
8	Gray	
9	White	

In the third band, silver indicates that the decimal point should precede the first digit. Gold indicates that the point lies between the first two digits. Other third band colors indicate the number of zeros to add.

add. A fourth band designates the manufacturing tolerance, and a fifth band (if it appears) specifies the reliability. Table 3.1 indicates the number each color represents. For instance, a resistor with bands of orange, white, yellow, and silver would have a **nominal*** value of 390 kilohms (or 390 K). The tolerance would be plus or minus 10%.

Remember from Section 2.5 of Chapter 2 that current in a resistance causes heating. For a given amount of current, the greater the resistance, the greater the heating will be. It is also true that for a given resistance an increased current will create more heat. Since the current through a fixed resistance is proportionate to the voltage applied, it follows that the heating effect for a given resistance is a product of the voltage across it and current flowing through it. To prevent resistors from burning up it is necessary to use those with physical size great enough to radiate away the heat developed. Resistors are rated according to the power (in **watts**) that they can handle, provided that they are well ventilated. Note the actual physical sizes for each rating as illustrated in Figure 3.3. (Power is more fully explained in Sections 10.8 and 10.9 of Chapter 10.)

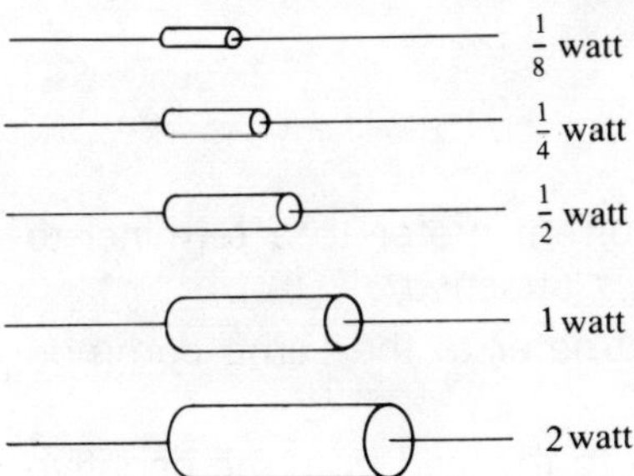

FIGURE 3.3 Actual size of resistors

Nominal means the intended value.

OHM'S LAW

Equipment needed

> 1—Current meter (10-mA maximum)
> 1—Voltmeter (25-V maximum)
> 1—3.3-K, 1/2-W resistor
> 1—4.7-K, 1/2-W resistor
> 1—33-K, 1/2-W resistor
> 1—25-V DC power supply
> 1—Breadboard

PROCEDURE

1. Connect the positive terminal of the current meter to the positive terminal of the power supply as in Figure A.

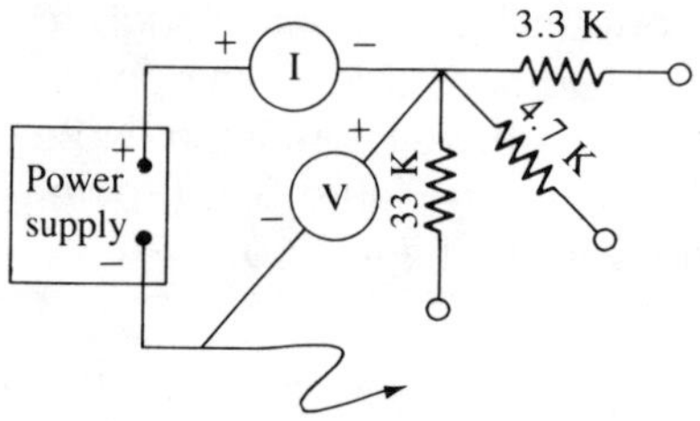

FIGURE A   Ohm's law demonstration

2. Connect the negative terminal of the current meter to a terminal to which one end of each of the resistors is attached.
3. Connect the positive terminal of the voltmeter to the same common terminal specified in step 2.
4. Attach the negative lead of the voltmeter to the lead from the negative terminal of the power supply.
5. Turn on the power supply and set the voltage to 25 volts.
6. Touch the negative voltmeter lead successively to the open end of each resistor and read the circuit current.

7. Repeat step 6 for voltages of 20V, 15V, 10V, 5V, and 0V.

8. Plot the data on a graph using current for a vertical axis versus voltage for the horizontal axis.

OBSERVATIONS

A. Note that the points lie on 3 straight lines.

B. Why isn't the current for the 3.3-K resistor exactly 10 times the current in the 33-K resistor?

VOLTAGE DROPS IN A CLOSED LOOP

Equipment needed

    1—Current meter (10-mA maximum)
    1—Voltmeter (25-V maximum)
    1—3.3-K, 1/2-W resistor
    1—4.7-K, 1/2-W resistor
    1—33-K, 1/2-W resistor
    1—25-V DC power supply
    1—Breadboard

PROCEDURE

1. Connect the resistors in series with the current meter and power supply as illustrated in Figure B.

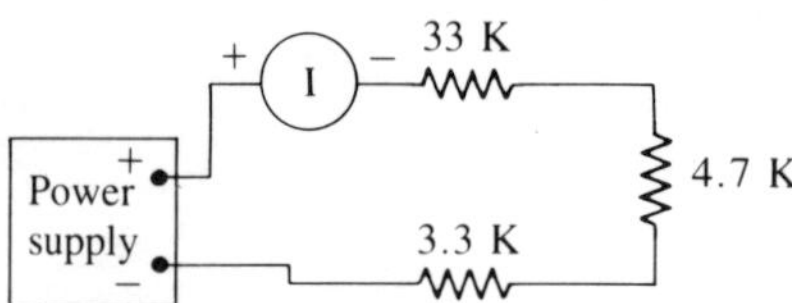

FIGURE B Voltage drop circuit

2. Turn on the power supply and adjust to 20 volts.

3. Measure the voltage across each resistor.

4. Note the current meter reading.

5. Move the current meter between the 33-K and 4.7-K resistor and note the reading.

6. Move the current meter between the 3.3-K resistor and the 4.7-K resistor and note the reading.

OBSERVATIONS

A. How closely did the three current readings agree? What does this tell you about the current in a series loop?

RELATIONSHIP OF VOLTAGE, CURRENT, AND IMPEDANCE

B. How closely does the sum of the voltage drops across the resistors agree with the power supply voltage? Why is there any disagreement?

QUESTIONS

From the following list select a symbol, letter, word, number, or unit to fill each of the blanks in the questions.

amperes	ohms	0.000302	17
d	power	0.050	28.8
decrease	*r*	0.15	30
divide	*t*	0.50	73
equality	tolerance	2.1	80
impedance	unchanged	12	84
increase	volts	14.5	300
multiply	*z*	16	4700

1. The general term for opposition to current is __________.

2. A 32-volt battery supplies 0.4 ampere of current in lighting a lamp. The resistance of the bulb is __________ ohms.

3. A resistance of 15 ohms has an EMF of 180 volts impressed across it. The current will be __________ amperes.

4. A voltage of __________ volts is required to obtain a current of 6 amperes through a resistance of 50 ohms.

5. Three battery cells of 1.5 volts each are placed in series aiding across a load of 30 ohms. The current to be expected is __________ amperes.

6. A formula states that $d = rt$. To manipulate this to solve for t in terms of d and r it is necessary to __________ both sides of the equal sign by __________.

7. A 1.25-ampere current results when a battery EMF of 36 volts is applied. The circuit resistance must be __________ ohms.

8. If the current in a 150-ohm resistance is 0.56 ampere, the voltage drop across the resistance will be __________ volts.

9. When the voltage supply to a circuit remains constant and the circuit resistance is reduced, the current will __________.

10. When the voltage across a circuit of fixed resistance is increased, the circuit current will __________.

11. If current through a resistance is decreased, the voltage across the resistance will __________.

12. A 45-volt battery supplies 1.5 amperes of current to a heating element. The element resistance is __________ ohms.

13. A 12-ohm instrument lampbulb is powered by a 6-volt battery. The current drain is __________ amperes.

14. The EMF required to cause a current of 0.025 ampere in a 680-ohm resistor is __________ volts.

15. A battery consisting of six identical cells in series has an output of 12.6 volts. The voltage of each cell is, therefore, __________ volts.

16. An electric iron draws 8.3 amperes when plugged into a 120-volt outlet. The resistance of the element is, therefore, __________ ohms.

17. A circuit contains a 47-kilohm resistor across which a voltmeter measures 14.2 volts. The current in the resistor is __________ amperes.

18. A carbon resistor has the following colored bands: yellow, violet, red, gold. The nominal resistance is __________ ohms.

19. Resistors of 60 ohms, 40 ohms, and 24 ohms when placed in parallel have a total resistance of __________ ohms.

20. A 20-ohm and an 80-ohm resistor are in parallel. Their total resistance is __________ ohms.

21. Resistors of 18 ohms, 22 ohms, and 33 ohms, when placed in series have a total resistance of __________ ohms.

22. The symbol used to represent impedance is __________.

23. When the quantities on each side of the equal sign of a formula are given the same mathematical treatment, the __________ still exists.

24. The physical size of a carbon resistor indicates the __________ rating.

25. The range within which a carbon resistor can be expected to deviate from the nominal value is called the __________.

4

Units,
Powers-of-Ten,
Scientific Notation

CONVENIENCE AND SIMPLIFICATION 4.1

The calculations performed so far have not really illustrated a need for a simplified method of expressing numbers. This is because in these problems, the numbers have been easy enough to handle using simple arithmetic. For instance, in problem 1 in Chapter 3, it was necessary to fit data into a formula as follows:

$$R = \frac{E}{I}$$

$$E = 32 \text{ V}$$

$$I = 0.4 \text{ A}$$

$$R = \frac{32}{0.4} = 80 \text{ }\Omega$$

Suppose the problem had indicated 32 V and 0.000040 A instead of the values given. Now the arithmetic becomes a bit more difficult.

43

$$R = \frac{E}{I}$$

$$R = \frac{32}{0.000040} = ?$$

Try this using conventional methods. If you found the correct answer you get an "A" for your effort, but also you must realize that there should be a simpler way.* There is. Of course, if you used your electronic pocket calculator, you found the answer with no trouble. Those little boxes are pretty smart! Now try this one.

$$R = 22,000,000 \ \Omega$$

$$I = 18 \ A$$

Find E $E = IR = 18(22,000,000) = ?$

Unless it was a fairly sophisticated model, the little box fumbled, didn't it? Yet calculations requiring nine digits or more are not uncommon in electronics. If your calculator did handle this, it probably used **scientific notation.** Scientific notation makes it possible to handle very large or very small amounts almost as easily as numbers between 1 and 10.

Units of measure have been devised on the basis of the metric system so that changes from a large unit to or from a smaller unit can be done just by moving a decimal point. Both of these simplifications depend upon a system called *powers-of-ten.*

4.2   *POWERS-OF-TEN*

When 10 is multiplied by 10, the operation can be indicated as 10^2. This is usually spoken of as "ten squared" because the area of a square that is 10 units on each side is obtained by multiplying the length by the width. In a similar fashion,

$$10 \times 10 \times 10 = 10^3 = 1000$$

10^3 is referred to as "ten cubed" because the volume of a cube that is 10 units on a side is the product of length, width, and depth. For the product of four or more 10's, the indicated product is called "ten to the fourth power" or "ten to the fifth power," etc. From the foregoing we can create a list such as Table 4.1.

---

*800,000 Ω

TABLE 4.1 Positive powers-of-ten

$$
\begin{aligned}
10 \times 10 \times 10 \times 10 \times 10 \times 10 &= 1{,}000{,}000 = 10^6 \\
10 \times 10 \times 10 \times 10 \times 10 &= 100{,}000 = 10^5 \\
10 \times 10 \times 10 \times 10 &= 10{,}000 = 10^4 \\
10 \times 10 \times 10 &= 1{,}000 = 10^3 \\
10 \times 10 &= 100 = 10^2
\end{aligned}
$$

Examination of the table shows that if the power-of-ten is reduced from 6 to 5, it is equivalent to having 1,000,000 divided by 10 to yield 100,000. Likewise, to reduce 10^5 to 10^4 we effectively divide by 10. This continues to apply going down the scale. Therefore, if 10^2 is reduced to 10^1 we divide 100 by 10 so that 10^1 is, obviously, 10. Now, reduce 10^1 to 10^0. What is the result now? If we follow the same pattern as before, 10 must be divided by 10 with the result shown that $10^0 = 1$. Again, if 1 is subtracted from the power-of-ten, $10^{-1} = 1/10 = 0.1$. A more complete table can be made as shown in Table 4.2.

This is a sort of shorthand for writing certain numbers. It would be valuable in itself, but its use can be extended. Consider a large number such as

$$62730$$

This is the same as 6.273×10000, but 10000 is 10^4. Therefore, our number can be expressed as

$$6.273 \times 10^4$$

Likewise, a very small number such as 0.00001478 can be written as

$$1.478 \times 10^{-5}$$

Thus, the technique of powers-of-ten is a short-form method of writing numbers.

TABLE 4.2 Extended powers-of-ten

$$
\begin{aligned}
1{,}000{,}000 &= 10^6 \\
100{,}000 &= 10^5 \\
10{,}000 &= 10^4 \\
1{,}000 &= 10^3 \\
100 &= 10^2 \\
10 &= 10^1 \\
1 &= 10^0 \\
0.1 &= 10^{-1} \\
0.01 &= 10^{-2} \\
0.001 &= 10^{-3} \\
0.0001 &= 10^{-4} \\
0.00001 &= 10^{-5} \\
0.000001 &= 10^{-6}
\end{aligned}
$$

4.3 *SCIENTIFIC NOTATION*

If numbers are written so that they are always greater than 1 but less than 10 and multiplied by a power-of-ten, they are expressed in scientific notation. Very large or very small numbers can be written with a minimum number of characters (a great advantage with pocket calculators with limited digits for read-out). In addition, calculations can be made quickly in your head. For instance, in the problem described in Section 4.1 in which EMF was 32 V and current was 0.000040 A, the solution becomes easier if the current is expressed in scientific notation. To do this, count the number of places the decimal point must be moved to the right to get a number between 1 and 10. In this case, a movement of 5 places is required. Since the move is to the right, a negative sign is placed in front of the 5, which then becomes the power-of-ten. We now have a number 4.0×10^{-5} that is exactly equivalent to 0.000040. The 32 V can now be divided by the 4.0 to yield 8.0. Next, this number must be divided by the 10^{-5}. If you like working with mathematics, you can easily prove that **the power-of-ten can be moved to the numerator if the sign of the power is changed.** So, in our example, the answer is that R equals 8.0×10^5, or 800,000 ohms.

The problem previously mentioned in Section 4.1 can now be solved with ease. Recall that the question was to multiply 18 by 22,000,000 using a simple four-function calculator. If the calculator produced an error or overflow signal, just express each number in scientific notation.* The 18 becomes 1.8×10^1 and the 22,000,000 is 2.2×10^7. So, now enter the 1.8 in the calculator, press $\times$, key in 2.2, and press =. The result is, of course, 3.96. The powers-of-ten can be done in your head. When 10^1 is multiplied by 10^7, the result is 10^8. Therefore, the answer to the problem is 3.96×10^8.

As calculations become more complicated, such as for electrical power (Chapter 10), the results displayed on the calculator will become less and less obvious. At that time the value of the use of scientific notation will be more apparent.

The real trick of using powers-of-ten, or the special case of it called *scientific notation*, is the manner of handling the powers in multiplication or division. The rule is that **when the powers-of-ten are**

---

*Any time your calculator registers an answer that starts with a decimal point followed by a string of zeros ending in a single digit, do not trust that answer. Convert the numbers in the problem to scientific notation; calculate an answer on the calculator ignoring the powers-of-ten; then calculate the powers-of-ten in your head and multiply times the calculator answer.

multiplied, the answer is ten with a power that is the sum of the powers. For instance,

$$10^3 \times 10^4 = 10^{(3+4)} = 10^7$$

or

$$10^{-2} \times 10^5 = 10^{(5-2)} = 10^3$$

The rule for division is just the opposite. Then the answer is a power-of-ten that is the difference of the divided powers. This is illustrated by the following:

$$10^6 \div 10^4 = 10^{(6-4)} = 10^2$$

or

$$10^4 \div 10^{-3} = 10^{[4-(-3)]} = 10^{4+3} = 10^7$$

In the last calculation it was necessary to subtract a negative three from the four. Subtraction of a negative number is equivalent to adding a positive. If this is difficult to understand, consider the following situation. Suppose you were in debt to a bank for $500. That would show on your statement as a negative number. If, unknown to you, that negative item was removed (subtraction) from your next statement, wouldn't that be the same thing as a gift (addition)? In short, just remember that *minus a minus is a plus!*

As has already been stated, when the numerator is divided by a power-of-ten in the denominator, it is equivalent to merely changing the sign of the power and placing it in the numerator. Likewise, of course, a power-of-ten can be moved to the denominator provided that you change the sign of the power.

It will be helpful, too, if you will remember that when expressing a number in scientific notation, you must note whether you have moved the decimal point to the right or to the left. When the point is moved to the right, the sign of the power-of-ten is negative ($-$). When it is moved to the left, the sign is positive ($+$).

METRIC SYSTEM UNITS 4.4

The metric system was devised to eliminate the difficult conversions such as we have all encountered when we needed to change from inches to feet, miles to yards, or other combinations. Nearly all of the nations of the world now use this system, and of the major powers only the United States has retained an archaic system.

TABLE **4.3** Abbreviations of prefixes designating power-of-ten multiple

tera	$= 10^{12}$	T	
giga	$= 10^{9}$	G	
mega	$= 10^{6}$	M	
kilo	$= 10^{3}$	k	*(centi* is not commonly
centi	$= 10^{-2}$	c	used with electronic
milli	$= 10^{-3}$	m	units)
micro	$= 10^{-6}$	μ	
nano	$= 10^{-9}$	n	
pico	$= 10^{-12}$	p	

In the metric system, the basic unit can be expressed with a prefix that designates a power-of-ten multiple. While additional prefixes were originally proposed, and a few more were added by Systeme International (SI) in 1960, the ones shown in Table 4.3 are most commonly used.

Using these prefixes we can express a quantity such as 14,000 volts as 14 kilovolts (abbreviated 14 kV). Or 0.073 ampere would be 73 milliamperes (abbreviated 73 mA).

Some peculiarities always seem to develop in any system, and the metric symbols are no exception. Capital letters are often used for the prefix symbols that represent positive powers-of-ten, and lower-case letters are used for the negative powers, but many people do not follow this practice. In fact, the system internationally used is a mixture, as shown in Table 4.3. This practice can cause confusion only between *mega* and *milli,* but since the values are a billion times removed from each other, the intended unit usually is obvious. All of the letters used are common to our alphabet except the μ for micro. This is the Greek letter mu (pronounced to rhyme with "new"). Since this. symbol is absent on most typewriters, the letter *u* is sometimes substituted.* The final oddity is the use of the letter *K* when indicating kilo-ohms. The Greek letter omega (Ω) is normally used to designate ohms, but when showing kilo-ohms it is most often omitted. As an example, 27,000 ohms is 27 kilo-ohms, or 27 kΩ. However, you will usually see it written 27 K. Again, there is little chance of confusion with this shortcut since the component being shown generally leaves ohms as the only logical unit to be considered.

In electronics, certain prefixes are most common with a small group of units. *Tera, giga, mega,* and *kilo* are used with the unit Hertz. *Mega* and *kilo* are most often used with ohms but occasionally will be applied to volts. *Milli* and *micro* are applied to volts, amperes, henries, and seconds. *Micro, nano,* and *pico* may be associated with farads or sec-

---

*Not a recommended practice, but the student should not be confused by it.

onds. (Farads and henries are the units of capacitance and inductance respectively, and will be discussed in Chapter 6.)

Custom usually dictates use of a convenient size of unit. Generally, this means use of no more than 3 digits to the right or left of the decimal point. For instance, 0.0012 A would be expressed as 1.2 mA. Similarly, 0.00075 V would be shown as 750 μV. An exception to this rule is common when expressing farads of capacitance when *pico* and *micro* are preferred over *nano*. Thus, 0.00000005 F would be expressed as 0.05 μF or 50000 pF but not as 50 nF.

QUESTIONS

From the following list select a word, set of numbers, or combination thereof to fill each of the blanks in the questions.

changed	sign	10^2	.0000417
five	six	10^3	.0000835
four	three	10^2	6280
milli	ten	10^3	16200
micro	two	10^4	62800
nano	10^{-4}	.0417	162000
negative	10^{-3}	.00301	628000
one	10^{-2}	.00417	1620000
pico	10^{-1}	.00835	6280000
positive		.0000301	
power			

In scientific notation:

1. 5280 is 5.28 × _________

2. 0.00138 is 1.38 × _________

3. 172.8 is 1.728 × _________

4. 77815 is 7.7815 × _________

5. 0.01982 is 1.982 × _________

6. When a number is converted to scientific notation the result is a number between _________ and _________ multiplied by a _________ of ten.

7. When a power-of-ten appears in a denominator, it can be moved to the numerator if the _________ of the power is _________ .

8. The metric prefix, *milli*, is equivalent to 10 to the _________ _________ power.

9. Ten raised to the power of negative nine is equivalent to metric prefix _________.

When converted to conventional form:

10. 8.35×10^{-5} = _________

11. 6.28×10^{3} = _________

12. 4.17×10^{-2} = _________

13. 1.62×10^{5} = _________

14. 3.01×10^{-3} = _________

15. When 10^{7} is multiplied by 10^{-4} the result is _________ .

UNITS, POWERS-OF-TEN, SCIENTIFIC NOTATION

5

Simple Amplification

USE OF AN AMPLIFIER 5.1

One of the major building blocks of electronics (and often the device that distinguishes the system as electronic rather than electrical) is the amplifier. To understand its use we will consider a typical public address system.

BLOCK DIAGRAM 5.2

A public address system consists of essentially three subsystems as shown in Figure 5.1. A microphone is a typical transducer. It converts sound waves into electrical impulses. Because these electrical signals are extremely weak they are fed into the **amplifier,** which uses the impulses as a pattern for

FIGURE 5.1   Public address system

controlling a larger source of energy so as to duplicate the input impulses' form in every way but with greater power. Thus, a magnified copy of the amplifier's **input signal** appears at its **output,** and is then fed to the loudspeaker. The loudspeaker is a transducer that resembles the microphone except that it does the opposite job. It converts the highly amplified electrical impulses back into sound waves. But, because the amplifier has added a great deal of power, the sound that entered the microphone can come from the loudspeaker much louder than it was originally. Also, because the amplified electric impulses can be carried by wires for considerable distances, the loudspeaker (or many loudspeakers) can be remotely located from the microphone.

5.3 *SOUND WAVES*

To understand truly the functions of the microphone and loudspeaker, it is necessary to examine the physics of sound. Sound can occur only where there is solid, liquid, or gaseous matter to carry it. Our most common experience with it is in air, which is a mixture of oxygen and nitrogen. Anything that can cause the air to be alternately compressed and rarefied (thinned) at a fairly rapid rate will create sound waves. As an example, the sketch in Figure 5.2 depicts a taut wire vibrating. As it moves to the right,

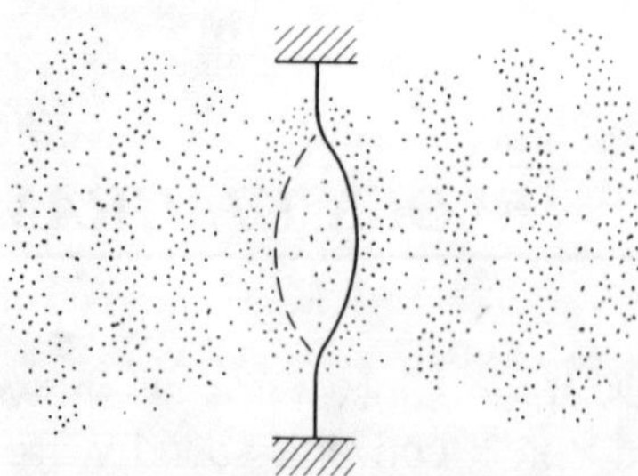

FIGURE 5.2   Sound waves from a vibrating wire

the air is compressed. This is symbolized by showing molecules of air (dots) being pushed together. As it moves to the left, the molecules are pulled farther apart. These alternate compressions and rarefactions travel away from the wire as sound waves. The formation and movement of the waves is similar to the formation of ripples a thrown pebble creates in a pond.

When sound waves reach a human ear they cause the ear drum to move in and out with the compressions and rarefactions. Through the intricate mechanism of the inner ear and the brain, we sense sound. A human ear is sensitive to vibrations from about 20 per second to about 20,000 per second.

MICROPHONE **5.4**

While all microphones convert sound waves into electrical impulses, there are several devices that can be employed to achieve this. One of the simplest is the **piezoelectric crystal.** Certain crystals such as quartz and rochelle salts, when properly cut, display a very unique property. When these salts are compressed or twisted, a voltage is momentarily generated between two facets* of the crystal. If the two facets are connected by a wire, a pulse of current will travel from one facet to the other. If the pressure or twist is removed, the crystal will snap back to its original shape and a pulse of current will flow back through the wire connecting the facets. The action of the crystal causes a voltage that creates a flow of electrons first in one direction and then in the opposite. Thus, the output is an alternating current.

A crystal microphone has a thin diaphragm that is vibrated by sound waves striking it. The diaphragm is attached to a piezoelectric crystal so that as it vibrates, the crystal is alternately twisted and released. The result is an alternating voltage that follows the pattern of sound waves striking the microphone. This voltage (signal) is extremely small, however, and generally must be amplified to be useful.

Another method of converting sound to an alternating current is the one used in early telephones. This consists of a package of carbon particles that, when squeezed by the sound pressure on a microphone diaphragm, has a lowered resistance. Conversely, when the carbon particles are decompressed by the opposite movement of the diaphragm, the resistance is increased. A battery placed in a series circuit containing the package (called a **carbon microphone**) will exhibit more current flow

---

*The crystals are cut in a manner similar to the cutting of a diamond so that there are flat surfaces (facets).

when it is compressed and less current when it is decompressed. What results is a direct current that varies up and down, just as would be the case if an alternating voltage were placed in series with a direct-current voltage. In fact, electronic receiving devices detect such a signal as a mixture of alternating and direct current.

5.5 *AC VOLTAGE AND MIXTURE WITH DC*

The foregoing will be clarified if direct current and alternating current are further defined.

When direct current is at a fixed value it is said to be in a **steady state.** A graph of steady-state, direct-current voltage shows a constant level of voltage over a period of time (illustrated in Figure 5.3).

Alternating current continually changes direction. This alternation occurs because the voltage polarity is repeatedly changed. The most common example of alternating current is a **sine wave.** Examples of this typical shape of electrical wave will be used in future illustrations. Figure 5.4 shows how the voltage polarity and voltage level may change over a period of time. Note that the maximum positive voltage (*a*) and the maximum negative voltage (*b*) are equal in *magnitude* but opposite in polarity (direction of current flow).

In Chapter 2 we established that voltages in series are added when aiding, or are subtracted when opposing. Consider the effect of an

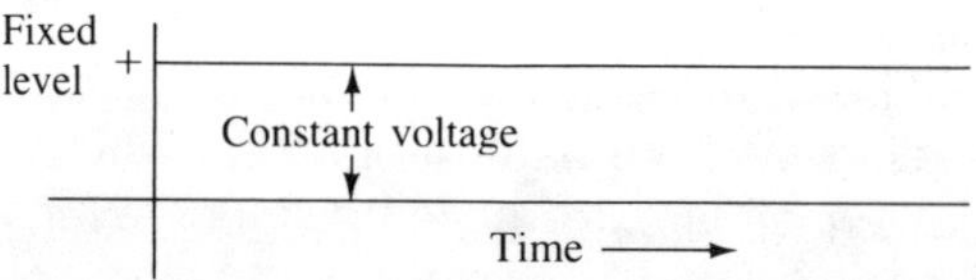

FIGURE 5.3   Graph of steady-state DC voltage

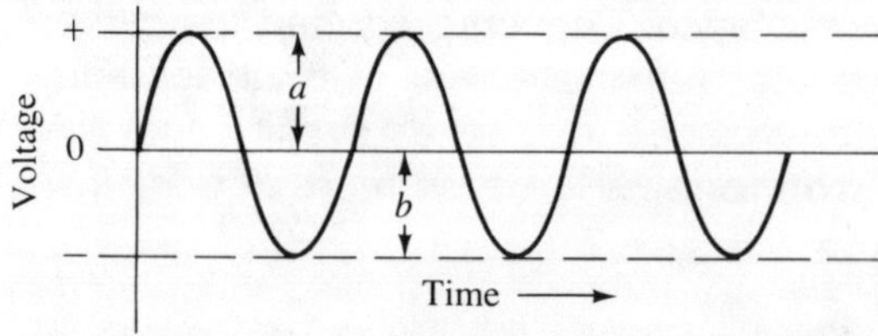

FIGURE 5.4   Alternating voltage sine wave

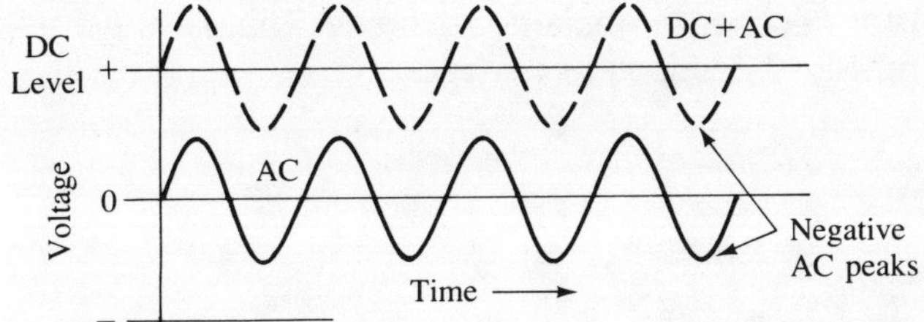

FIGURE 5.5 Sum of steady-state DC and an AC sine wave

alternating-voltage generator in series with a battery. Part of the time the two are in series aiding, and part of the time they are opposed. This can be graphed as seen in Figure 5.5.

In the illustrated example, the DC level is greater than the negative **AC peak.** The dashed line indicates that the sum of the DC and AC results in a constant DC polarity with a varying amplitude. This waveform has the same appearance as the signal coming from the carbon microphone described in Section 5.4. Thus, a varying DC signal appears as a mixture of AC and DC.

FREQUENCY 5.6

When an alternating signal is referred to as *steady state,* the term means that the maximum and minimum voltage swings, as well as the time between these alternations, have become constants. The common household electric outlet is a good example. The power company continuously supplies about 115 volts at a **frequency** of 60 **Hertz.** A *Hertz* is a frequency unit of one **cycle** per second. A cycle is a series of events that repeat themselves. In Figure 5.4, for example, three cycles of an alternating voltage are pictured. Each cycle contains a positive polarity followed by a negative polarity. The first half of the cycle has a voltage buildup to a maximum and a gradual decline, as illustrated in Figure 5.6. When the voltage

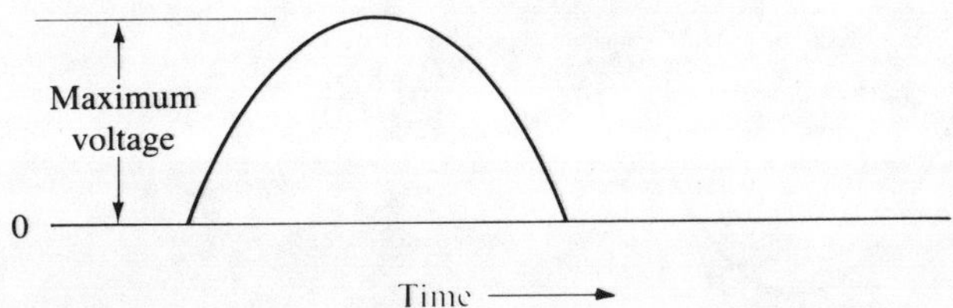

FIGURE 5.6 One-half cycle of a sine wave

reaches the zero point after decline, the polarity reverses. That is, the voltage begins to drive current in the opposite direction.

5.7 *PERIOD AND WAVELENGTH*

The graph in Figure 5.7 shows one complete cycle. The time required to complete a full cycle is called the **period.** Usually time is represented in formulas by the lowercase letter t. The length of one period of time, however, is represented by the capital T.

As mentioned earlier, frequency is measured in Hertz. If a cycle requires one-hundredth of a second, its perod (T) is 0.01, and its frequency (f) is 100 Hertz. Therefore,

$$T = \frac{1}{f} \quad \text{or} \quad f = \frac{1}{T}$$

A common term associated with frequency is **wavelength.** This is the length in space of one complete cycle. Electromagnetic waves travel through space at a speed of 186,000 miles per second. If this quantity is measured in metric units, the speed is 300,000,000 meters per second.

If we could see a 186,000-Hertz wave as it was beamed out into space for just one second, there would be 186,000 cycles, one after the other, and each would be one mile long. On the other hand, if the frequency were 30 Hertz, there would be 30 cycles and each would be 6200 miles long. It follows that wavelength is easily found by dividing the speed by the frequency. Wavelength is usually represented by the Greek letter λ (lambda); therefore,

$$\lambda \text{ (meters)} = \frac{300,000,000 \text{ (meters per second)}}{f \text{ (Hertz)}}$$

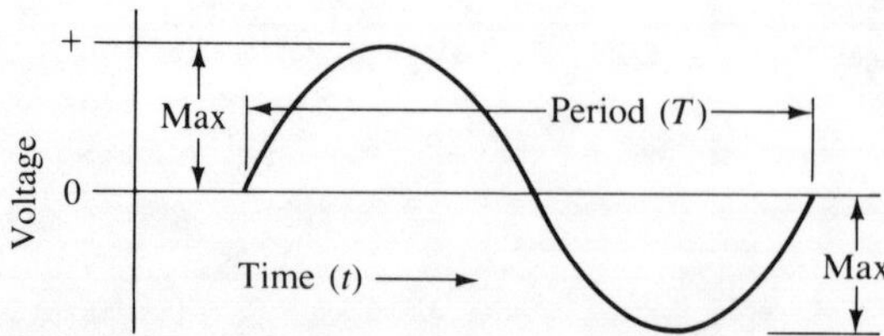

FIGURE 5.7 Single cycle of a sine wave

Of course, a little manipulation of this formula as explained in Section 3.4 gives us a formula for frequency in terms of speed and wavelength:

$$f = \frac{300 \times 10^6}{\lambda}$$

Alternating-current voltages can be produced with frequencies of billions of Hertz, but one unique frequency deserves special attention. When a current has a frequency of zero, so that alternation does not occur at all, it is simply direct current.

AMPLIFIERS **5.8**

In the public address system we have been investigating, the microphone converted sound waves into an AC voltage or a mixture of DC and AC voltages. In either case, the AC portion carried the electrical equivalent of the sound wave. These AC voltages are characteristically very small and would be useless if they were not increased to greater **amplitudes** (made stronger). This is the function of an amplifier.

Two **active devices** are currently used to provide amplification. Historically, the first was the **vacuum tube.** While tubes are still employed in a few applications, the **transistor** began to displace it in the 1950's. In more recent years, the transistor (as a discrete device) has been replaced increasingly by **integrated circuits.** These devices incorporate entire circuits of transistors, diodes, resistors, and associated components into packages so small that the circuits are visible only under a microscope (see Chapter 15).

BIPOLAR TRANSISTORS **5.9**

Two basic types of transistors are now common. The earliest type developed was the **bipolar** transistor. This name is applied because two types of **current carriers** are employed.

Up to now, we have considered only one type of current carrier, the electron. Electron flow can take place because in some materials (such as copper and silver) electrons in the outermost orbital shells of the atoms are very weakly attracted to the nucleus of the atom. These electrons can be detached by small forces. In fact, just the heat present at room temperature is enough to cause many of them to leave their parent atoms and fly randomly about the surrounding space like cub scouts on their first camping trip. The wanderers are called **free electrons.**

In certain crystalline **semiconductor** substances, notably **silicon** and **germanium**, a different type of movement can take place. In these materials, a few atoms of a different element are placed in the pure crystal. These "impurities" are selected to have fewer electrons in their outermost orbitals than are in the outer shells of the overwhelming majority of surrounding atoms. The material is then called *P* (*positive*)-*type.* In crystal structures the atoms arrange themselves in a definite pattern so that the forces between the atoms are balanced and symmetrical. The strange atom in such a structure leaves a **hole.** That is, there is a place (hole) where an electron would be if an atom of the majority element were there instead of an atom of the impurity. Any electron from an adjacent atom or a free electron injected into such a crystal can alight temporarily in one of these holes. When the electron moves to fill another hole, the former hole appears again. Now it may seem odd at first, but it is convenient to consider this a movement of the second hole to the position of the first rather than as a movement of the electron.

If the preceding idea seems unclear, consider this analogy. Suppose a vertical tube is nearly filled with oil and plugged at both ends. If the tube is turned upside down, a bubble of air will rise from the bottom to the top. Yet, it is just as reasonable to say that the oil moved to the bottom.

Since we consider the hole to move, this type of current is called **hole flow,** and the holes are called **current carriers.**

Semiconductor substances are also modified by causing a few atoms (per million) to have more electrons in their outermost orbitals than do the atoms of the major material. This addition lowers the resistance by supplying electrons that can be easily broken free. Such material is called *N* (*negative*)-*type.*

In the bipolar transistor, both types of current carriers are employed: electrons and holes. The essential elements of this transistor are the **emitter, base,** and **collector.** Symbolically, such transistors are shown illustrated in Figure 5.8.

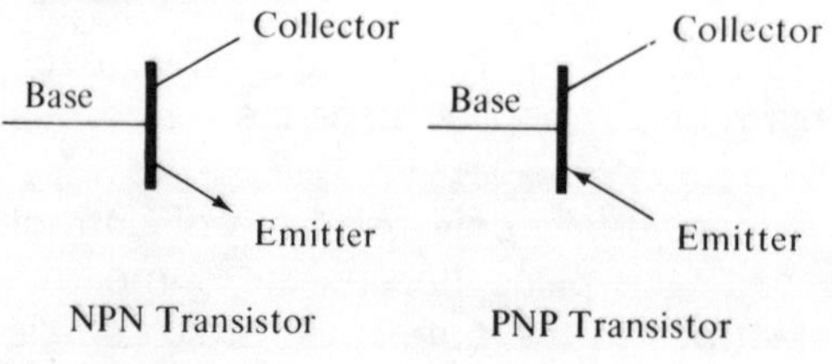

FIGURE 5.8 Symbols for bipolar transistors

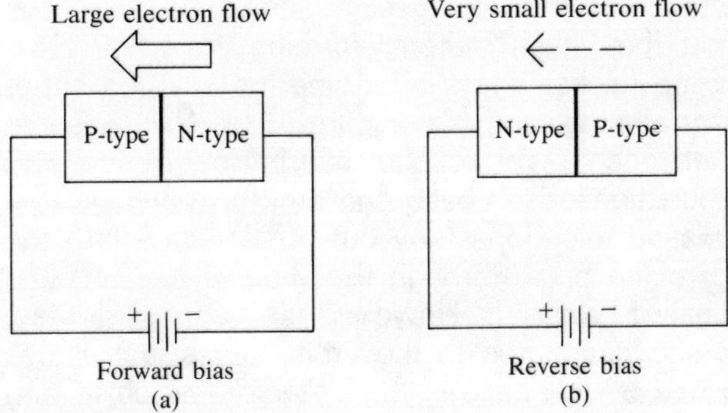

FIGURE 5.9 PN junction diode

The direction of the arrow representing the emitter in Figure 5.8 indicates whether the transistor is an NPN type or a PNP type. These designations describe the construction of the transistor and tell us how to connect it properly. A junction of P-type and N-type material presents a greatly different opposition (impedance) to current depending on the polarity of voltage applied. Such a **PN junction** is called a **diode.** If the P-type material is made more positive than the N-type, the condition is **forward bias** and the impedance is small (Figure 5.9a). On the other hand, if the N-type material is more positive than the P-type, the condition is termed **reverse bias,** and the impedance is very large (Figure 5.9b).

Suppose, for instance, that we consider an NPN transistor in the circuit shown in Figure 5.10. Battery A, the variable resistor, the base, and the emitter are a closed loop. Note that the negative terminal of battery A is connected to the emitter, which is N-type material. The base of P-type material is, therefore, more positive (so the base emitter junction is forward biased). Battery B makes a closed loop with the load resistor and the whole transistor. However, note that the positive terminal of the battery is connected to the N-type material of the collector (so the base/collector junction is reverse biased).

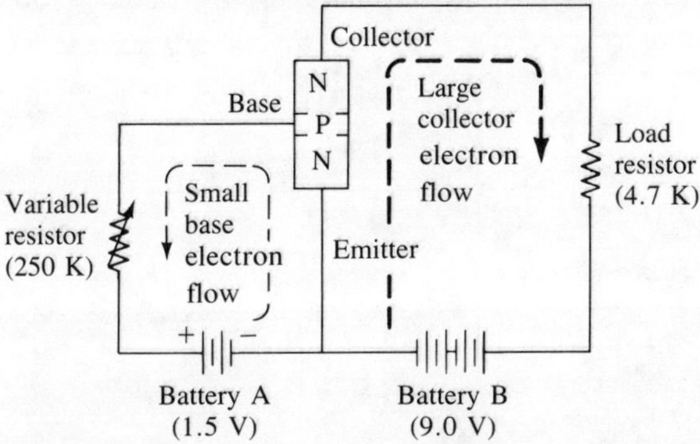

FIGURE 5.10 Forward-biased NPN transistor

To understand the functioning of a (NPN) transistor, it is first necessary to note that the P-type material forming the base is extremely thin compared to the thicker layers of N-type material (on either side) comprising the emitter and collector (see Figure 5.10). Also, observe that battery A is typically of much lower voltage than battery B.

Since the junction of the base and emitter materials (the **PN junction**) is forward biased, electrons flow out of battery A into the emitter, and are attracted to the base (through the variable resistor) and back to the positive terminal of battery B. However, the negative contact of the larger battery B is *also* connected to the emitter, and just across the very thin "fence" formed by the base is the collector, which in turn is connected (through the load resistor) to the positive side of battery B. So, even though the collector is N-type material, it is connected to a much higher *positive* potential (+9.0 V) than the base circuit can supply (+1.5 V). There are too few holes in the thin P-type base material for all those free electrons in the thick N-type emitter. *But,* the positive potential from battery B is close at hand in the collector.

What happens is that most of the electrons in the emitter "jump the fence" into the collector, where they flow through the load resistor into battery B. Thus, the more positive the base is with respect to the emitter (the more electrons it attracts), the more electrons there are available to "jump" into the collector. In effect, a small forward flow from emitter to base controls a much larger flow from emitter to collector. And, if the base current is cut off, so (effectively) is the flow to the collector. This explanation is greatly simplified from the actual atomic physics of transistor action, but it is still a close enough approximation to be quite useful.

From the foregoing it is obvious that what the base current is doing is varying the resistance to current between the emitter and collector. Indeed, that function is where the device got its name; the name *transfer resistor* was simply shortened to *transistor*.

It is rather inconvenient to have a circuit that requires two batteries as in Figure 5.10, so a modification such as shown in Figure 5.11 is more common. In this circuit, the voltage supplied to the base is ob-

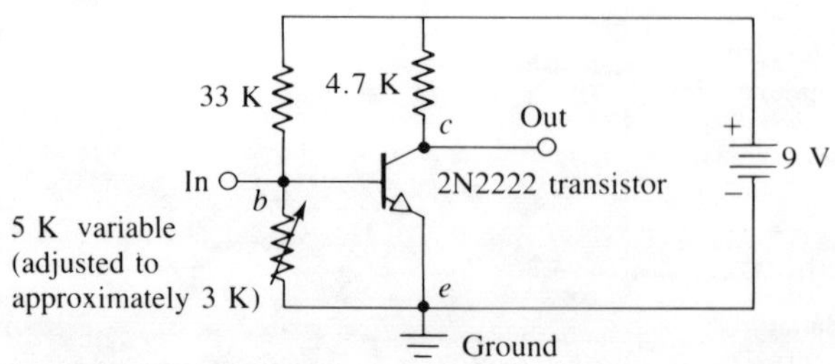

FIGURE 5.11 Simplified amplifier circuit

tained by a **voltage divider.** Observe that when 9 volts is placed across a circuit consisting of a 33-K resistor in series with a 3-K resistor, the current will be 250 microamperes:

$$\frac{9}{33000 + 3000} = .00025 \text{ ampere}$$

This amount of current will cause a voltage drop across the 3-K resistor of about 0.7 volt.

$$.00025 (3000) = 0.75 \text{ volt}$$

Thus, the 9.0 V is *divided* by the two resistors.

Only 0.6 to 0.7 volt is necessary to forward bias successfully the base-emitter junction of the transistor.* Therefore, the voltage divider removes the necessity of a second battery. We can place a 5-K **potentiometer** (variable resistor) in series with the 33-K resistor and adjust the voltage division at the base so that the electron flow from emitter to collector is neither fully turned on nor fully turned off. Suppose the base voltage is set so that the voltage at the collector (point *c*) is 4.5 volts. This will occur when the base current is such that it causes a flow from emitter to collector that will cause a voltage drop of 4.5 volts across the load resistor (4.7 K). A microphone can be connected with one lead to point *b* and the other to point *e*. When the microphone output causes additional current to enter at *b*, the increased base-emitter current causes an increase in collector current (from *e* to *c*). The current increase through the 4.7-K resistor will cause the voltage drop across it to increase. The upper end of the 4.7-K resistor is tied to the battery and is, therefore, locked to +9.0 volts. Consequently its lower end must drop. Actually, as previously explained, the transistor is acting as a variable resistor, forming a voltage divider with the 4.7-K resistor. (See Figure 5.12.)

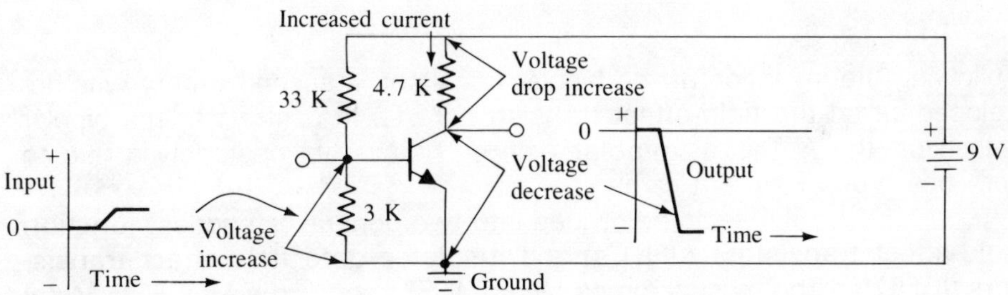

FIGURE 5.12   Input voltage increase causing output decrease

---

*0.6 to 0.7 volt is normal for silicon transistors. Approximately 0.2 volt is enough to forward bias germanium transistors.

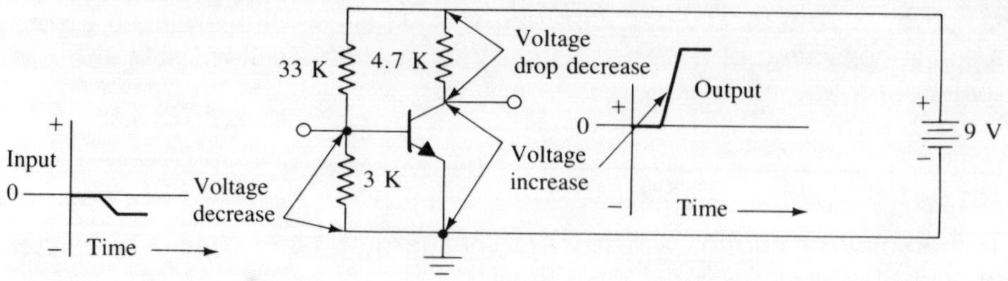

FIGURE 5.13   Input voltage decrease causing output increase

Conversely, when the input to the base decreases the forward bias, the collector current is reduced, which causes less voltage drop to occur across the 4.7-K load resistor. Again, because the upper end of the resistor is locked to +9 volts (or 9 volts above **ground***), the voltage at the lower end (at the collector) must rise. (See Figure 5.13.)

With this circuit in actual test, using a 1000-Hz input signal alternating from 5 mV negative to 5 mV positive, the voltage at point *c* of Figure 5.12 alternated (increased and decreased) by 500 mV from the 4.5-volt DC level. Voltage amplification is the output change divided by the input change. In this case the **voltage gain** was

$$\frac{500}{5} = 100$$

5.10 *FIELD-EFFECT TRANSISTORS*

Following the invention of the bipolar transistor, a second variety was developed called the **field-effect transistor (FET)**. In contrast to the bipolar transistor, the FET is a **unipolar** device. That is, its operation is due to only one type of current carrier.

FET's can be divided into two general categories: **junction field-effect transistors (JFET's)** and **insulated-gate field-effect transistors (IGFET's)**. The most common type of IGFET is called a **MOSFET** (**M**etal **O**xide **S**ilicon FET).

**Ground* is a term frequently used to designate the common or negative connection in electronic circuits (and in some devices it is actually physically connected to the earth).

All FET's have three terminals, but their names are different from those given to bipolar devices. Instead, they are called **source, gate,** and **drain.**

FET's (like bipolar transistors) are actually variable resistors. Instead of varying the resistance by control of the base current, the FET is controlled by the gate voltage.

The JFET is made by surrounding a channel made of one type of material with a blanket of another type. For instance, a core of N-type material would be wrapped with a blanket of P-type material, as shown in Figure 5.14a. If voltages are placed on the terminals as shown, the junction between the P-type and N-type material is always reverse biased. In this case the P-type material is biased more negatively than the voltage applied to the source. Such a condition is necessary to operate this type of FET (N-channel). As the amount of reverse bias is increased, the electrons in the N-type material near the junction with the P-type section are repelled. As they move away, the area they left begins to have so many holes that it appears as P-type material. The result is that the junction effectively moves inward and causes the channel to be restricted. This effect is called **depletion.**

The voltage applied between the source and drain will cause current in inverse proportion to resistance (Ohm's law). However, the resistance is determined by the cross-sectional area through which the current carriers can flow. Since application of more voltage to the gate has the effect of narrowing the path of current from source to drain, current is reduced. This effect is similar to the idea illustrated in Figure 5.14b, where a rubber tube carrying air or water can be pinched by the fingers to restrict the flow.

Application of an alternating voltage to the gate will alternately relieve and increase the pinching effect. As a result, the flow from source to drain will vary according to the variations of the signal. Although the gate signal may be very weak, if the current is high between source and drain, great amplification can be achieved by causing large changes in the voltage drop across a load resistor in series with the FET.

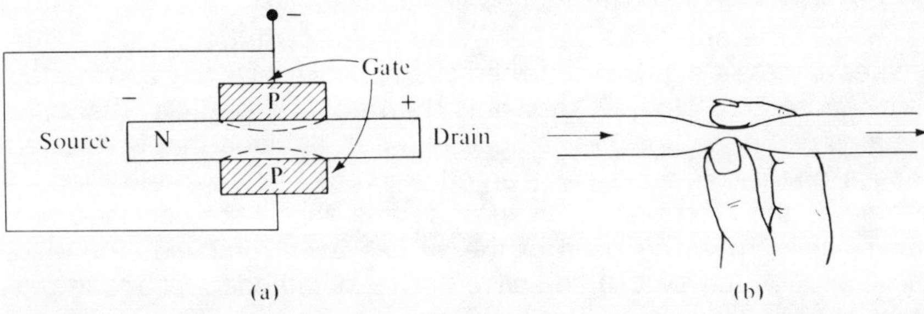

FIGURE 5.14 Pinching effect of JFET

As the name suggests, an insulated-gate FET is constructed with insulation between the source-to-drain channel and the gate material. This removes the necessity that the gate-to-channel junction be reverse biased. As a consequence, such FET's are not limited to the depletion mode of operation as described for the JFET. The channel of the IGFET can be enlarged as well as pinched. This additional property is called **enhancement.**

5.11 *INTEGRATED CIRCUITS*

Most modern circuits contain one or more **integrated circuits (IC's).** These extremely compact circuits, which have been developed since the 1960's, are made up of many diodes, transistors, resistors, and associated circuit elements. With the exception of a hybrid type IC that has some discrete components mounted on it, all the components in an IC are manufactured simultaneously on a tiny slice of semiconductor material.

5.12 *VACUUM TUBES*

Vacuum tubes were the first amplifying devices, and they are still used at present in very high power transmitters and other devices subjected to high temperatures. Their principles of operation are still very important today because a special type of vacuum tube, the **cathode-ray tube,** is used as the picture tube for televisions, as the screen for radar and oscilloscopes, and as the monitor for computer terminals.

The heart of a vacuum tube is the **cathode.** This is a surface of special metals that is maintained at a negative potential and heated to emit electrons by a process known as **thermionic emission.** The emitted electrons gather around the cathode in a sort of cloud that is referred to as **space charge.** A positively charged element (metal plate) called an **anode** attracts the electrons. Both cathode and anode are enclosed in a glass envelope from which most of the air has been removed. The electrons flow through the evacuated space between cathode and anode provided that no other forces are present. One or more **grids** (metal screens) may be placed between the cathode and anode. If the grids are negatively charged, they impede the electron flow. If they are positively charged, the flow is accelerated. For more details, see Section 16.4.

STAGING 5.13

In general, each active device that forms an amplifier with its associated components is called a **stage.** Sometimes the output signal of one amplifier stage is used for the input signal of a second amplifier stage. With either transistors or vacuum tubes, when only one device is used, the amount of amplification may not be sufficient. In some cases, the voltage may have been increased enough, but not the available power. For example, a power stage may be necessary to drive a loudspeaker to the volume needed. Either voltage or power (or both) can be increased by multiple stages.

Figure 5.11 illustrates a single amplifier stage. A second stage would have a similar configuration and would receive its input from the output indicated in the illustration.

LOUDSPEAKER 5.14

The normal loudspeaker used today makes use of two forms of magnetism to convert the electrical energy from the amplifier into the sound it produces.

A magnet must have two poles, an N pole and an S pole. There is a force between these poles called **magnetomotive force,** which is usually abbreviated **MMF.** The MMF exists as a **field** just as we speak of a gravitational field. **Field** is a term applied to forces of attraction or repulsion that exist between two or more locations. A field cannot be seen but can be detected and measured. It is frequently convenient to think of the magnetic field as **lines of force** like curved wire strands connecting two opposite magnetic poles, but it must be realized that these lines are imaginary (see Chapter 1, Experiment 1).

The familiar horseshoe magnet is a variety of **permanent** magnet. Permanent magnets are so called because they retain their magnetism for a long time. This is a characteristic called **retentivity.** Most permanent magnets are made of iron, but two other metals, nickel and cobalt, also are magnetic. Certain alloys are now manufactured that make better magnets than the elementary metals. Generally speaking, hard materials make better permanent magnets than soft ones. This is due to the fact that magnetizing a piece of material results in aligning the molecules of the material. In hard materials, once they are aligned to form a magnet, they do not readily get shaken into misalignment again.

Temporary magnets can be made of such materials as soft iron. These materials are easily magnetized, but when they are removed from the magnetizing field the molecules quickly fall into misalignment and the material loses its magnetism.

In most loudspeakers a magnet of the permanent variety is used. All speakers used today are of this type, although this was not always so. Hence you may occasionally see that they are called P.M. speakers, indicating permanent magnet.

In addition to the permanent magnet, modern speakers use **electromagnetism.** In the 1800's, Hans Christian Oersted discovered that when a current flows in a conductor, a magnetic field surrounds it. André-Marie Ampere found that coiling a conductor of current produced the equivalent of a bar magnet, except that he could control the strength of the magnetic field by changing the amount of current permitted to flow.

As stated previously, magnets have polarity that we label N or S. The polarity of an electromagnet is determined by the direction of the current flow. This principle is used in a loudspeaker in the following way.

A permanent magnet is mounted on the frame of a speaker at the rear center, as in Figure 5.15. A stiff paper cone is attached to the outer ring of the frame, and a coil of wire is attached to the cone's center so that the coil surrounds but does not touch the permanent magnet. If a small current is allowed to flow through the coil in one direction, the coil and paper cone are drawn back toward the permanent magnet. This occurs because the coil's electromagnetic field is polarized (momentarily) opposite that of the permanent magnet, and unlike poles attract. If the current is reversed, the magnetic polarity of the coil is reversed and the coil and cone are repelled. If these reversals are done rapidly, the cone vibrates at a frequency that can be heard. The output from an amplifier is coupled to the loudspeaker coil so that the speaker cone is forced to vibrate at the rate originally injected as a signal from the microphone into the amplifier.

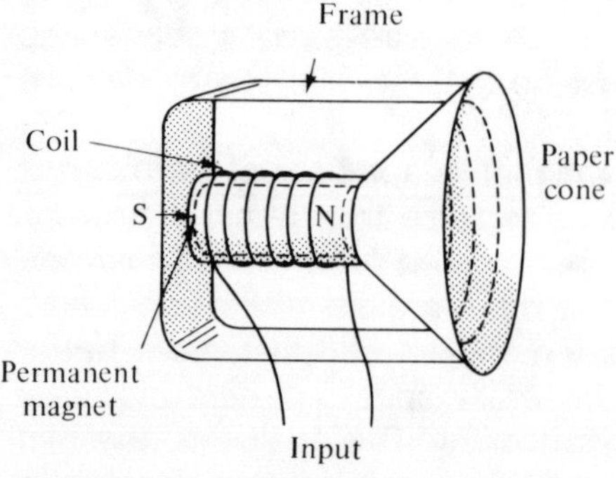

FIGURE 5.15 Loudspeaker

SUMMARY 5.15

What can be seen, then, from the preceding discussion is that sound waves enter a microphone that converts this mechanical energy to an alternating voltage. Since this voltage is very weak it must be amplified, which consists of using this weak signal to control a larger current from another source. A similar action occurs in an automobile where small, easy movements of a foot on the accelerator are ''amplified'' to control many horsepower. In our PA system the amplified alternating voltage is finally applied to the electromagnet of the loudspeaker. This time the electrical energy is converted to sound that closely approximates the original sound entering the microphone, except in magnitude.

BIPOLAR TRANSISTOR AMPLIFICATION

Equipment Needed

- 1—NPN silicon transistor (2N2222 or other)
- 1—2.2-K, 1/2-W carbon resistor
- 1—10-K, 1/2-W carbon resistor
- 1—22-K, 1/2-W carbon resistor
- 1—33-K, 1/2-W carbon resistor
- 2—9-V batteries
- 2—Battery clips
- 2—1 1/2-V dry cells
- 1—Double cell holder
- 1—Breadboard
- 1—Voltage meter

PROCEDURE

1. Construct the transistor circuit illustrated in Figure A. The 18 volts can be from a DC power supply or from the 9-volt batteries in series. Nearly any NPN silicon transistor will work in this circuit.
2. Place two 1.5-V dry cells in series and connect the most negative to ground. (The symbol for ground is ⏚.)
3. Connect the negative terminal of a meter to ground.
4. Measure the voltage from the base to ground.
5. Measure the voltage from the collector to ground.

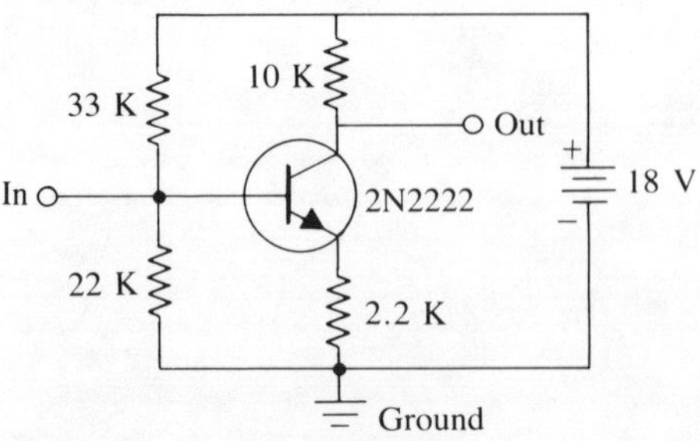

FIGURE A Amplifier circuit

6. Touch the positive lead from the two 1.5-V cells to the base and measure the collector voltage and the base voltage.

7. Calculate the change in collector voltage from step 5 to step 6.

8. Calculate the change in base voltage from step 4 to step 6.

9. Calculate the ratio. The number from step 7 is divided by the number from step 8 to arrive at the voltage gain.

10. Reverse the input voltage: Connect the negative terminal of the two 1.5-V cells to the base and the positive terminal to ground.

11. Measure the voltages at the base and collector.

12. Subtract the reading made in step 4 from the one made of the base in step 11.

13. Subtract the reading made in step 5 from the one made at the collector in step 11.

14. Calculate the voltage gain and compare it with the result obtained in step 9.

ELECTROMAGNETIC REPULSION AND ATTRACTION

Equipment Needed

    1—Bar magnet
    1—20-foot length of #28 varnished wire
    2—1 1/2-V dry cells
    1—Double cell holder
    1—120-V to 6-V stepdown transformer*
    4—22-ohm, 1/2-W carbon resistors
    1—Loudspeaker (3 inch or larger)
    1—Ohmmeter
    1—Breadboard

Safety Note

This experiment calls for use of a transformer connected to the 120-V power outlet. Be sure that the power cord is not frayed, that the connections of the cord to the primary winding are well covered, and that the metal housing is connected to ground.

PROCEDURE

1. Remove the varnish insulation from about 1 inch of each end of the 20-foot length of varnished wire.
2. Wrap a coil of wire around the bar magnet loose enough so that it can easily slide over the magnet. Leave 5 inches of lead on each end. (This is easy to do if a 4 × 4-inch strip of paper is folded twice so that it is 4 plies, 1 inch wide, 4 inches long. Wrap the paper around the magnet. Wind the turns over the paper (2 layers deep, 1 inch long), tape the turns with transparent tape, slide the paper and coil off the magnet, remove the paper, and return the coil to the magnet.)
3. Connect the leads from the coil to the breadboard.

---

*Any low-voltage (2-V to 12-V) transformer will work, but the series resistors may need to be changed accordingly.

    SIMPLE AMPLIFICATION

4. Connect the 1 1/2-volt cells in series and momentarily apply voltage across the coil. (It will be necessary to support the magnet so that the coil is free to move.)

5. Reverse the battery leads and observe the movement again.

6. Wrap your left hand around the coil so that the fingers point in the direction of the electron flow in the coils. Your thumb will be pointing in the direction of the N magnetic pole of the coil.

7. Set the ohmmeter on the $R \times 1$ scale. While touching one terminal of the speaker with one probe from the meter, gently touch or lightly scratch the other probe across the opposite speaker terminal. Observe that the cone of the speaker moves when the second terminal is touched. If the movement is not visible you should at least be able to hear its effect. (Set the ohmmeter on an $R \times 100$ or larger scale and try again.)

8. Reverse the ohmmeter probes and observe that the speaker cone moves in the opposite direction.

9. Connect the four resistors in series on the breadboard, and connect the secondary of the transformer across them.

10. Attach one lead of the speaker to one of the transformer secondary terminals.

11. Touch the other speaker lead to the connection between the first two resistors from the connection made in step 10. Observe the action of the speaker. If the tone from the speaker is not loud enough, move the speaker lead to the next resistor junction. Avoid making the output too loud or the speaker could be ruined.

QUESTIONS

From the following list, select a symbol, letter, word, set of numbers, or combination thereof to fill each of the blanks in the questions.

AC	gate	piezoelectric	.0025
amplifier	hard	PNP	.025
base	holes	resistance	.25
circuit	IGFET	retentivity	1.95
collector	increase	seconds	2.0
cone	integrated	silicon	2.5
DC	JFET	soft	3.06
decrease	junction	source	20
depletion	microphone	speaker	100
drain	micro	staging	200
electrons	MOSFET	thermionic	300
emitter	nano	tube	400
enhancement	NPN	vacuum	500
fields	period	zero	20000

1. The subsystems of a public address system are __________, __________, and __________.

2. Sound can be transmitted through anything but a __________.

3. The range of normal human hearing is from __________ Hertz to __________ Hertz.

4. When a momentary voltage is caused by distortion of a material, it is called a __________ effect.

5. Compressing and stretching a package of carbon particles in the carbon microphone causes a change in __________ of the package.

6. When current from a battery is caused to vary it is often considered to be a mixture of __________ and __________.

7. If a wave has a frequency of 400 Hertz, the time for one cycle is __________ second.

8. The time for one cycle is called the __________.

9. A radio wave with a frequency of 98 megahertz has a wavelength of __________ meters.

10. Sound travels at about 1000 feet per second. The wavelength of 512 cycles per second (middle C on the piano) would be __________ feet.

11. The period of a radio wave whose wavelength is 150 meters is __________ __________ __________.

12. If a radio wave had a wavelength of 1 meter, the frequency would be __________ megahertz.

13. The frequency of a direct-current voltage is __________.

14. Current is the rate of movement of current carriers. In semiconductors these carriers are __________ and __________.

15. The three terminals of a bipolar transistor are the __________, __________, and __________.

16. For proper operation of a bipolar transistor, the junction of the __________ and __________ must be forward biased, and the junction of the __________ and __________ must be reverse biased.

17. If base current of a transistor is increased, one should expect the collector current to __________.

18. An increase in collector current causes a (an) __________ in the voltage drop between the collector and ground.

19. The three terminals of an FET are __________, __________, and __________.

20. The symbol illustrated in Figure B is for a/an __________ bipolar transistor.

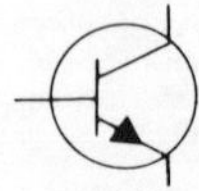

FIGURE B Transistor symbol

  SIMPLE AMPLIFICATION

21. The J in JFET stands for the word __________.

22. The most common type of IGFET is the __________.

23. The action that causes the channel of an FET to have greater resistance is called __________.

24. The action that reduces the resistance of the channel of an FET is called __________.

25. Variations in the current through the coil in a loudspeaker cause the __________ to be vibrated so that sound is produced.

26. When many diodes, resistors, transistors, and other components are developed as a circuit on a tiny slice of semiconductor material, the product is called an __________ __________.

27. The earliest amplifying device was the __________ __________.

28. The existence of invisible forces such as gravitational, electrostatic, and magnetic are spoken of as __________.

29. Permanent magnets are made from __________ materials, while temporary magnets are made from __________ materials.

30. The creation of a space charge above the heated cathode of a vacuum tube is due to __________ emission.

31. If the amplification of one active device is insufficient, two or more may be used so that the output of one becomes the input to the next. This technique is called __________.

32. The ability of a magnet to remain magnetized over some time is called __________.

6

A Radio Transmitter

For most people the subject of electronics immediately suggests radio and television. This is only natural because these communication media have been two of the greatest commercial developments of the past fifty years. Today, even though such things as lasers, computers, and robots have captured the spotlight of attention, the importance of radio and TV has increased. It is appropriate, therefore, to gain some understanding of their operation.

The block diagram of a simple radio transmission system is shown in Figure 6.1. To understand the operation of this system, it is necessary to comprehend the manner in which wireless transmission occurs. Two related phenomena occur: One is magnetic, and the other is electrical. Together they are called **electromagnetics**.

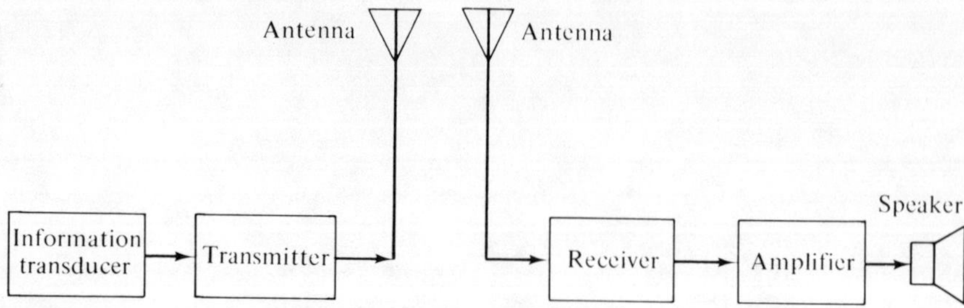

FIGURE 6.1 Radio transmission system

6.3 *ELECTROMAGNETIC FIELD*

In Section 1.8 it was established that like charges repel each other, but unlike charges attract. The force between two charges is spoken of as a **field**. If the charges are slowly reduced or increased, the field will be reduced or increased, but no useful **electromagnetic wave** will be generated. If the increase and decrease are made very rapidly, the field cannot collapse as quickly as the charge decreases, and that field is "pushed away" by the next increasing field. An analogy to this can be seen in the generation of sound waves as explained in Section 5.3 with Figure 5.2. If the wire in the figure is slowly pushed from one side to the other while gripped in the fingers, the compression and rarefaction of the air are too slight to produce sound, because the air can slowly drift out of the way of the wire and drift back in following it. When the wire is allowed to vibrate, however, the air cannot flow so rapidly and a *wave* of compressions and rarefactions is formed. Actually, any movement of the wire will cause some slight wave, just as will any change in a charge. However, just as vibrations of fewer than 20 cycles per second are not considered sound, so electromagnetic waves with a frequency of fewer than several thousand Hertz are not classified as **radio waves**.

A **broadcast antenna** provides the mechanism to accomplish what has just been described. An alternating voltage can be applied to a length of conductor (antenna) causing its two ends to alternate in polarity. Alternating currents flow up and down the conductor producing a field that expands and collapses at such a rapid rate that a wave is driven away from the antenna. When this wave strikes another (receiving) antenna some distance away, the changing field causes an alternating voltage between its ends with the same frequency as that applied to the broadcast antenna. The wave is called *electromagnetic* since it contains electrical and magnetic components.

Again, an analogy may be helpful in getting a feeling of what happens. Imagine a small pond with two corks floating in it. If one of the corks is struck with a stick so that it bobs up and down, ripples will gradually spread out from the cork and will eventually cause the other cork also to bob up and down. The cork struck with the stick is equivalent to a broadcasting antenna, and the other cork is like a receiving antenna.

To achieve the transmission of a radio wave there must be a source of high-frequency AC voltage and current. This need is met at the broadcast station by a device called an **oscillator**.

OSCILLATION 6.4

Oscillation is one of the most common processes in nature. For instance, the vibration of a tightly stretched guitar string or a leaf in the breeze, or the swinging of a pendulum are all forms of oscillation. Any time an object is twisted, compressed, or stretched and then allowed to seek freely a condition where it is stable, it will go through a series of oscillations. Consider a child's swing, for example. When the swing is pushed and then left to return to its original stationary position it does not stop immediately, but swings to and fro past that rest point until all the energy expended in the original push has been dissipated by friction. While it is swinging, it is said to oscillate. In effect, the energy of the push is stored momentarily. The energy stored is the work required to push the seat of the swing against earth's gravitational pull to a position higher than when it was at rest. Most of this potential energy is converted to momentum when the seat is allowed to swing back toward the positon of rest. As the seat passes its rest point the remaining momentum is once again converted to potential energy. Thus, the seat oscillates back and forth.

In electronics, as in the above example, there must be some provision for the momentary storage of potential energy if oscillation is to occur. Two devices display this capability. One is called a **capacitor**, and the other is termed an **inductor**. With these elements it is easy to make an electronic oscillator.

CAPACITANCE 6.5

If two conducting surfaces are placed near (but not touching) each other, and electrons are driven onto one of the surfaces, electrons on the other

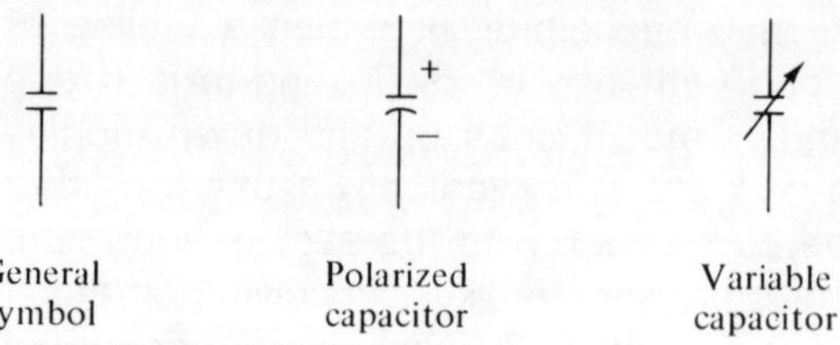

FIGURE 6.2 Capacitor symbols

surface strain to move away. (*Like charges repel.* See Section 1.8.) In practice, the two conductors (called **plates**) are insulated from each other, either by air or by some more compact insulator like plastic film, glass, or mica. One plate of the **capacitor** is momentarily connected to the negative terminal of a voltage source, while the other plate is connected to the positive terminal. Electrons flowing onto the first plate repel free electrons in the second plate, which are then drained into the positive-voltage terminal. If the voltage source is disconnected, a charge remains between the two plates. If a conductor is connected then between the **charged** plates, current will flow until the voltage between them is equalized, or the capacitor is **discharged**. A quantity of electrons can, therefore, be stored in this manner. The greater the voltage applied to a capacitor the greater the number of charges that can be stored. Since the amount of charge that can be stored in a given capacitor depends upon the applied voltage, the unit of measure of capacitance is based upon the amount of charge that can be stored per volt.

Capacitance is measured in a unit named for Michael Faraday* called the **farad**. *One farad of capacitance will store one coulomb of charge with an EMF of one volt.* This is an extremely large unit for normal electronics use. Most capacitors are measured in microfarads (μF or mfd), meaning one-millionth of a farad. Even this unit is frequently too large and picofarad (pF) is used. This is one-millionth of one-millionth of a farad.

Capacitors are specialized components and will be discussed in greater detail in Chapter 13. The symbols used for them are shown in Figure 6.2.

6.6 *INDUCTANCE*

In Section 5.14, when the action of a loudspeaker was discussed, it was observed that when a current flows through a conductor a magnetic field

---

*One of the most important of the early scientists, Faraday developed much of what is known about electricity.

is established around it. If the flow is direct current in a steady state, the magnetic field is also steady. In this condition the field is said to be stationary. That is, at any particular distance from the conductor the strength of the field remains constant. It will be strongest close to the conductor and weaker farther away.

If the current is increased in the conductor, the field at each point becomes stronger. The field has the appearance of moving outward from the conductor since the **field strength** that was observed close to the conductor can now be found farther out. In an opposite way, if the current is decreased, the field appears to collapse back into the conductor. The strange part of this, however, is that when voltage is applied to increase the current, there is delay as the current accelerates to a higher amount. Also, when the voltage is decreased, it takes time for the current to reduce. This might be very difficult to understand if it were not for another rather startling fact. If a second conductor is placed beside the first one, a current will flow in the second whenever the intensity of the magnetic field around it changes. The direction of flow in the second conductor is opposite to that of the current in the first conductor that generated the increasing or decreasing field. So, an increasing field around the original conductor will generate a voltage to oppose the increasing current. The reverse action occurs with decreasing current.

In the preceding discussion, observe that a voltage was **induced** in a conductor lying alongside the original without being connected. This is called **mutual inductance**. As the two conductors are moved apart, less of the magnetic field overlaps both, and the induced voltage falls off very rapidly. The effect of the field upon the flow of current in the original conductor is called **self-inductance**. This amounts to a temporary storage of energy in the magnetic field. A straight wire is not very efficient as an inductor, so it is customary to wind it into a coil so that the magnetic field from each turn of the coil crosses over the other turns (see Figure 6.3). This greatly increases the inductance. Inductance can be increased, also, by providing an iron core for the coil to improve the path for the magnetic field.

Since the inductive effect affects the *rate* at which current will increase or decrease following a *change* in applied voltage, the unit of inductance, called the **henry (H)**, is defined in terms of time. A 1.0-henry inductor will produce 1.0 volt of **back EMF** (or **counter EMF**) if the current is increased (or decreased) at a rate of 1.0 ampere per second.

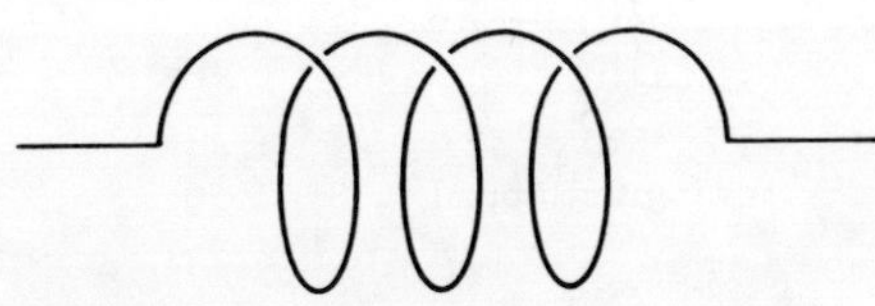

FIGURE 6.3 Inductance coil

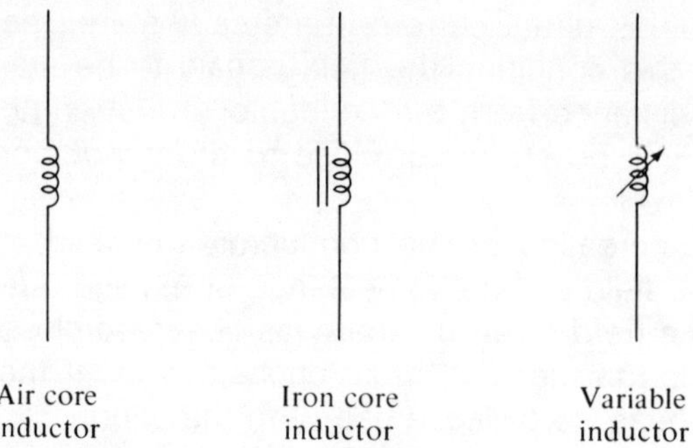

FIGURE 6.4 Inductor symbols

Inductance is analogous to the **inertia** of **mass**. For instance, if it is necessary to push an automobile it is not a good idea to get a running start. A bloody nose would most likely result! The car opposes a sudden change in speed because of its mass. Once the vehicle is moving, it would be foolish to try to stop it by putting a foot in front of a wheel. Again its mass or inertia opposes any change in speed. Similarly, an inductor opposes any increase or decrease in the current flowing through it. The larger the inductance the greater the opposition to change. Symbols for inductors are shown in Figure 6.4.

6.7 *TRANSFORMER*

Mutual inductance is the principle behind the operation of the **transformer**. (Typical transformer symbols are illustrated in Figure 6.5.) If two independent coils are wound around the same piece of soft iron, a change in current in one of them will cause a changing magnetic field to exist around the other winding. A voltage is thereby created in the other winding that will cause current to flow in it if the winding's two ends are connected to a closed circuit. In this manner currents will flow in both windings even though they are electrically isolated from each other. We will examine

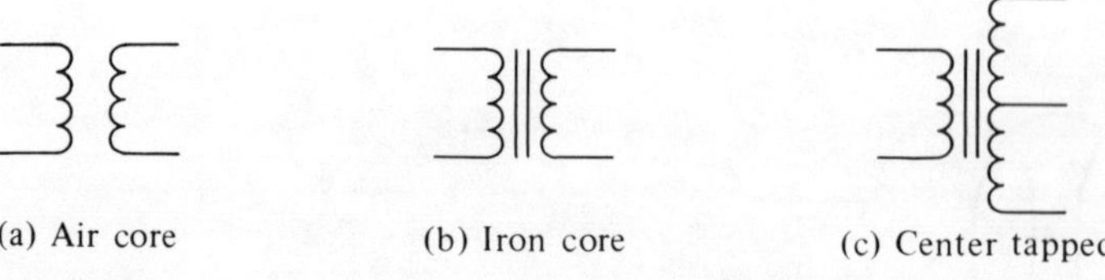

FIGURE 6.5 Transformer symbols

transformers more thoroughly in Chapter 12. For now it is sufficient to recognize that the winding into which a signal enters is called the **primary winding**. The one where output is taken is the **secondary winding**.

TANK CIRCUIT **6.8**

A capacitor and inductor provide the components for an electronic mechanism for oscillation. The capacitor can store energy in the difference of charge on its opposing plates, and the inductor can store energy in the magnetic field surrounding its conductor. When placed in a closed loop as shown in Figure 6.6, they form what is commonly known as a **tank circuit**. Assume that by some outside source of energy, the capacitor can be given a charge so that plate A has a greater number of electrons than plate B. Immediately, the electrons will attempt to flow to B through point C, the coil, and point D. This will cause a magnetic field to expand around the coil as the current begins to flow. There will, of course, be a back EMF created in the coil to oppose this flow, so its increase will be gradual. Furthermore, as the flow proceeds, the difference in potential between plates A and B will decrease. These two effects working together cause the *rate of increase* of current to begin to decline, so the magnetic field will begin to collapse. Now, instead of opposing the current flow, the collapsing field begins to aid the failing voltage from the capacitor. Even when the capacitor voltage falls to zero the collapsing field continues to drive electrons from A to B until B becomes charged as high as A was originally. That is, this happens unless there is some resistance in the circuit. If there is resistance, the voltage of B will be a bit less.

Since the capacitor has been charged once again and the magnetic field has completely collapsed, the electrons on B will start to flow back to A through the coil and the whole process described above

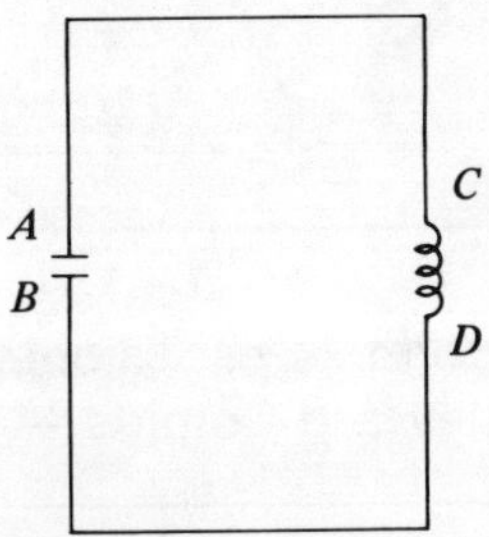

FIGURE 6.6 Tank circuit

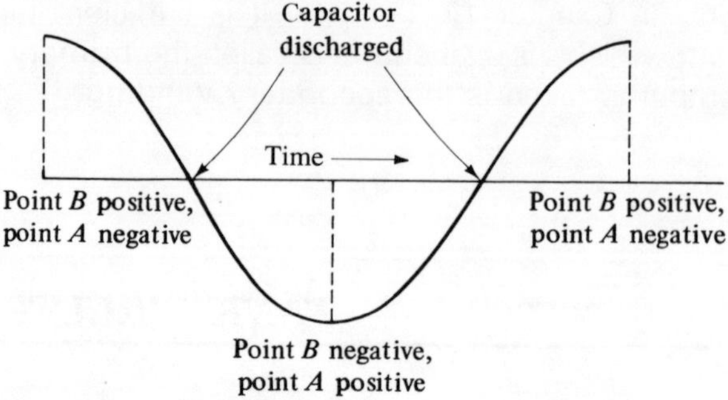

FIGURE 6.7 Voltage change in a tank circuit

will occur again (but with the current in the opposite direction). The full cycle of change of voltage at point *B* is shown in Figure 6.7.

The operation described will repeat over and over until all the energy is dissipated in whatever resistance is present. (Under laboratory conditions, with practically zero resistance, a tank circuit has been observed to oscillate for over two years.) The waveform created in a tank circuit is called a **sine wave**. A sine wave is the most common form of oscillation in nature and one that receives a large amount of attention in electronics.

The speed of oscillation depends upon the product of the values of capacitance and inductance used. The frequency of oscillation is called the **resonant frequency**.

Resistance in the circuit will soon cause the amplitude of the oscillations to decrease to a negligible amount, just as a child's swing eventually comes to rest. The pendulum of a clock would do the same thing if it were not for a small amount of energy supplied to it by the clock spring at just the right moment in its swing. So what is needed is an electronic device that will give a small impulse to the circuit every cycle at just the right instant. Any device that amplifies can be used for this purpose. Part of the output of the device (such as a transistor or tube) can be "fed back" to give the input to the tank circuit an extra nudge to keep it oscillating.*

6.9 *OSCILLATOR*

A large number of circuits have been designed to perform as oscillators. Two of the most elementary types are known as **Hartley** and **Colpitts**. A

---

*This feedback is so easily done that it often occurs with amplifiers when it is undesired. One of the frustrations frequently encountered by technicians is to keep an amplifier amplifying instead of oscillating.

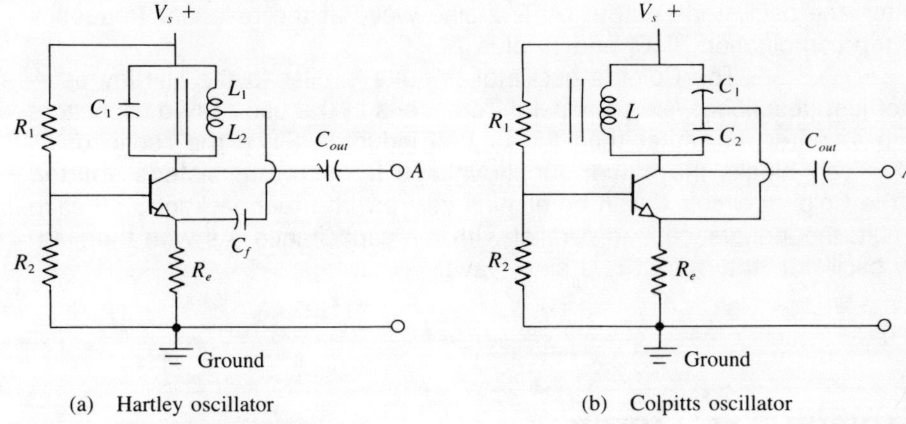

(a) Hartley oscillator   (b) Colpitts oscillator

FIGURE 6.8 Typical oscillators

typical Hartley oscillator using a transistor is shown in Figure 6.8a, and a similar Colpitts oscillator is shown in Figure 6.8b.

Let us first consider the operation of the Hartley. Then the manner in which the Colpitts operates can be easily inferred.

As soon as the voltage source V_s in Figure 6.8a is applied, the voltage division provided by R_1 and R_2 causes the transistor to be forward biased. Electrons flow from ground through R_e and the transistor, and through L_1 and L_2, to V_s. The immediate increase in current through the inductor (L_1 and L_2 in series) is opposed by self-induced back EMF. Capacitor C_f is connected to the junction of L_1 and L_2 (called a *tap*) so that *changes* in the voltage at that point (AC currents)—*but not* the power supply voltage from V_s (DC current)—are coupled to the transistor's emitter. At first, as the current through the transistor and L_1 and L_2 is increasing rapidly, the voltage coupled back via C_f makes the emitter become more negative with respect to the base. This increases the current through the NPN transistor, and the further increase is again transferred through C_f (and so on) until more forward bias will no longer increase collector current. At this point the "inertia" of the inductor causes current to continue to flow and charge C_1; however, the *change* is now toward lesser current. This opposite change is coupled to the emitter by C_f. Also, C_1 eventually reaches its maximum charge, and then begins to discharge through L_1 and L_2. The combination of these changes coupled through C_f drives the emitter positive with respect to the base, and turns off the transistor. Now C_1 completes its discharge through L_1 and L_2, and the inductance once again causes the current to continue flowing, reversing the charge on C_1. Notice, this is an oscillating tank circuit as explained in Section 6.8. Actually, the **feedback** through C_f from the tank circuit causes the transistor to turn on momentarily at just the right time in the cycle of oscillation. This gives the tank the boost of current needed to overcome resistance in its circuits and continue steady oscillation. As it is the tank circuit that controls the transistor, the oscillator's output (A) is a sine wave at the resonant frequency of the combination of C_1 and L_1 plus L_2.

The Colpitts oscillator is quite similar to the Hartley oscillator just described. The principal difference is in the use of two capacitors with a center tap rather than the tapped inductor. Since the capacitor C_1 effectively blocks the power supply voltage from the transistor's emitter, in the Colpitts circuit C_f can be eliminated from the feedback loop. In each circuit, the inductance is in parallel with the capacitance. As with the Hartley oscillator, the output is a sine wave.

6.10 *BUFFER AMPLIFIER*

It is common practice to "feed" the output of an oscillator (called a **carrier wave**) into an amplifier called a **buffer**. This is done to isolate the oscillator's output from large power drains (called **loads**) since **loading** would change the oscillator frequency. Also, the additional stage is necessary to amplify the tiny oscillator signal. The output of the buffer amplifier is then fed to a **power amplifier** that is connected to a broadcast antenna. The power amplifier greatly boosts the output of the buffer to give the transmitter a useful broadcasting range.

6.11 *TRANSMITTING INFORMATION*

The signal produced by the oscillator and amplifier so far described is of a constant frequency and amplitude. Such a signal at a receiver has very limited value because it indicates only that the transmitter and receiver are working. No other information is conveyed. The simplest way to improve this situation is to place a switch or key at the transmitter so that the signal can be switched off and on in a code, such as the familiar Morse code used in telegraphy. However, a different method must be used to transmit the voice of a singer, the music of a rock 'n' roll band, or a symphony.

Two kinds of information—*volume* and *frequency*—must be broadcast simultaneously to reproduce with fidelity a signal as complex as music. The remainder of this chapter explains two means of transmitting such information. Note how these two methods handle the problem differently.

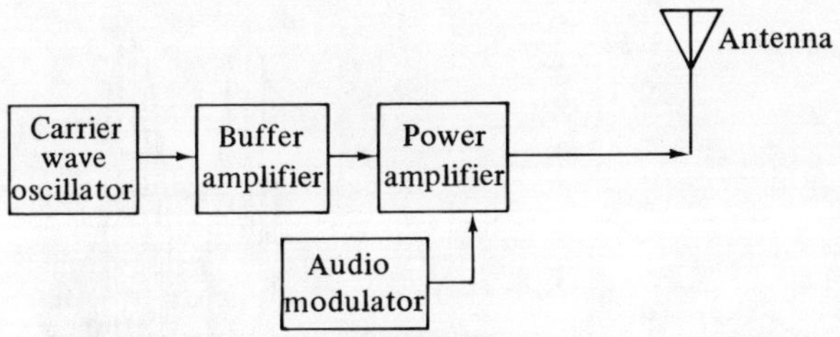

FIGURE 6.9   Carrier wave modulation

MODULATION **6.12**

To accomplish voice and music transmission, a technique called **modulation** is employed. The two most common types of modulation are **amplitude modulation (AM)** and **frequency modulation (FM)**. Modulation means simply *impressing a carrier wave with the information to be transmitted.*

Amplitude modulation is produced in a transmitter by causing the **audio** (speech or music) frequencies to control the voltage (amplitude) and current of the signal going from the power amplifier of a transmitter to the antenna (see Figure 6.9). Several methods can be employed, but one of the easiest to understand is known as **collector modulation**.

COLLECTOR MODULATION **6.13**

Suppose the output of a microphone or other audio pickup is amplified and the output of the audio amplifier is coupled to the collector circuit of a **radio-frequency** (R-F) power amplifier so that the collector voltage is increased or decreased by the audio signal. The output of the power amplifier is affected so that its amplitude is large when the audio signal is positive and small when the audio signal is negative.

Figure 6.10 is a diagram of a simplified circuit performing the functions of the power amplifier and audio modulator blocks shown in Figure 6.9. Transistor Q (the power amplifier) receives the carrier-wave frequency (via T_1) from the oscillator and buffer and amplifies it. However,

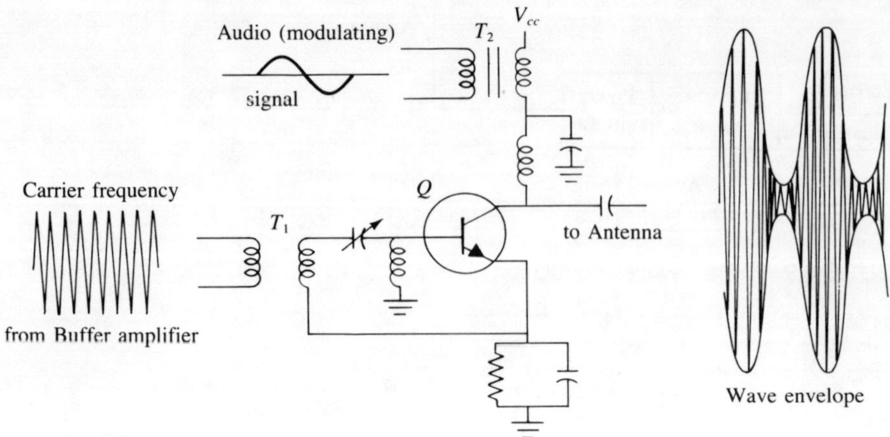

FIGURE 6.10   Amplitude modulation

the amount of output will depend upon the transistor's collector voltage, which originates from the DC voltage source V_{cc}.

An amplified audio signal originating from a microphone (or other transducer) flows in the primary of transformer T_2, which induces a changing voltage in the secondary. This voltage adds and subtracts from V_{cc} causing the collector voltage of Q to vary. Changing the voltage on the collector thus causes the amplification of the carrier wave to vary at the frequency of the modulating audio signal.

In Figure 6.10 an outline has been drawn following the positive and negative peak values of the modulated wave. This is commonly referred to as the wave **envelope**. It is significant that the top half of the envelope is a facsimile of the audio signal and the bottom half is its mirror image. Thus, the carrier-wave envelope is the receptacle for the audio information the carrier wave is carrying.

Observe that the *volume* of the audio signal is represented in the modulated wave by the *amount* the output is changed in amplitude. The *frequency* of the audio signal is represented in the modulated wave by the *frequency* at which the output is changed in amplitude.

6.14   *FREQUENCY MODULATED TRANSMISSION*

One of the problems with amplitude modulation is that disturbances such as lightning and the arcing of motor brushes tend to cause changes in the amplitude of the carrier wave. This is received as **static**. One means for eliminating static interference is to use a different method of transmission called **frequency modulation (FM)**.

In an FM modulator the modulating audio wave causes the carrier wave to be changed in frequency, but not in amplitude. One way to achieve this is to have a microphone that will change the value of capacitance in the tank circuit of the oscillator. Figure 6.11 illustrates such a circuit. Note that the microphone diaphragm also forms a movable plate that constitutes one-half of the tank-circuit capacitance. As the diaphragm vibrates closer to the stationary plate, the capacitance is increased, which causes the tank's resonant frequency to be reduced. As the diaphragm vibrates back, the distance between the capacitor's plates is increased, thus lowering (temporarily) the capacitance in the tank circuit. With lower capacitance the capacitor requires less time to charge and discharge each cycle, so the period is reduced and the frequency is increased. In this way the carrier wave changes in *frequency* in step with the sound striking the microphone, while the carrier wave *amplitude is constant* since the collector voltage does not vary. The resulting wave forms are illustrated in Figure 6.12.

MODULATION SUMMARY 6.15

Section 6.11 explained that both volume and frequency information are required for voice or music transmission. Table 6.1 summarizes how these two are handled differently by amplitude or frequency modulation.

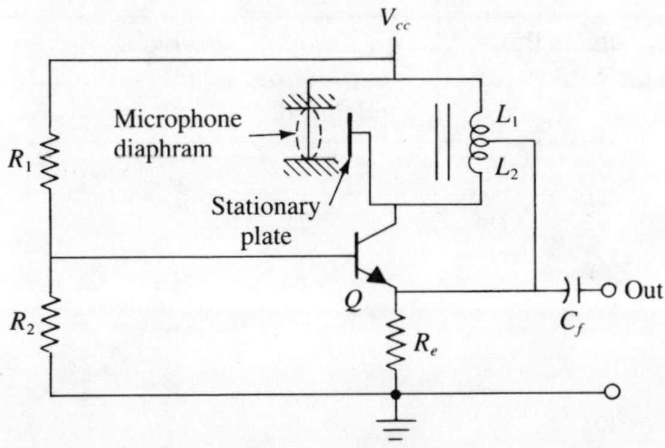

FIGURE 6.11 Frequency modulation

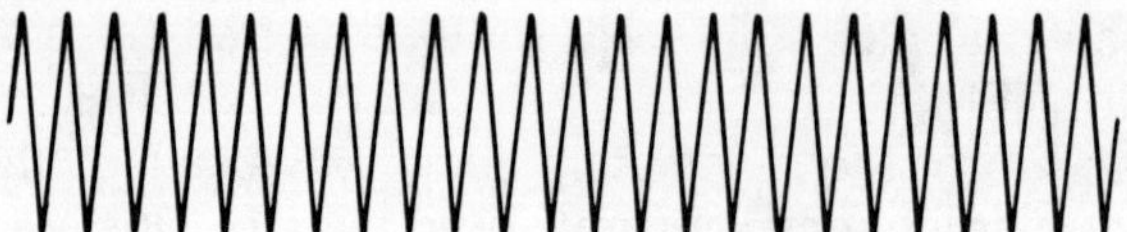

Carrier wave when microphone diaphragm
is not vibrated

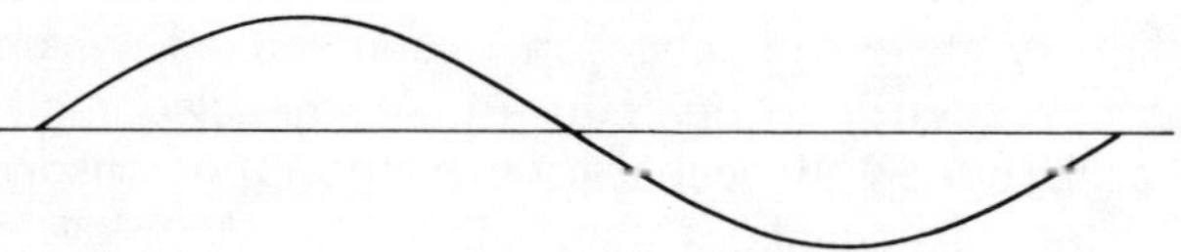

Audio waveform of air striking microphone

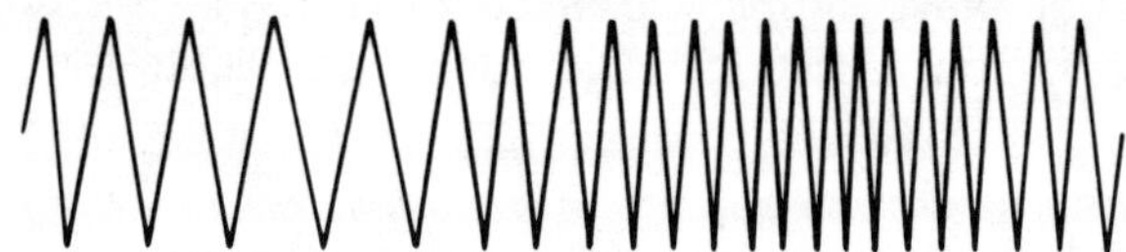

Frequency modulated wave

FIGURE 6.12 Frequency modulation of waveform

TABLE 6.1 Comparison of AM and FM Modulation

Information to be trans- mitted	Method of transmission	
	AM	FM
Frequency	The *frequency* at which the carrier *amplitude* is changed.	The *frequency* at which the carrier *frequency* is changed.
Amplitude	The *amount* by which the carrier *amplitude* is changed.	The *amount* by which the carrier *frequency* is changed.

BUILD AND TEST AN OSCILLATOR

Equipment Needed*

- 1—2N2222 transistor
- 1—150-mH inductor
- 1—3.9-microfarad capacitor
- 1—2-microfarad capacitor
- 1—0.47-microfarad capacitor
- 1—4.7-K, 1/2-W carbon resistor
- 1—10-K, 1/2-W carbon resistor
- 1—33-K, 1/2-W carbon resistor
- 1—Breadboard
- 1—Oscilloscope
- 1—20-V DC power supply

PROCEDURE

1. Construct the circuit illustrated in Figure A.
2. Connect the output of the oscillator at point *A* to the vertical input of

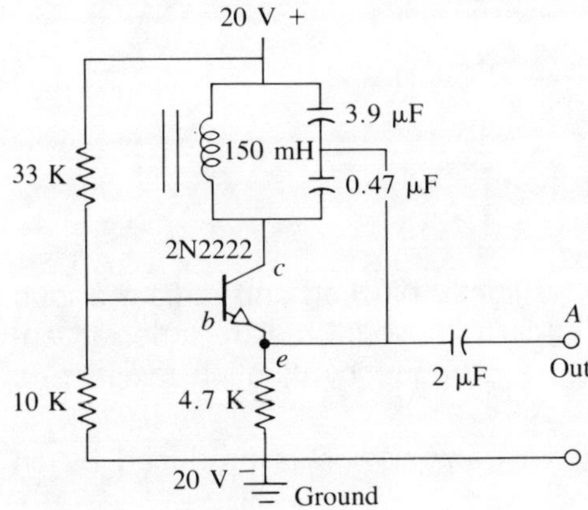

FIGURE A Colpitts oscillator

---

*The capacitor and inductor values are not critical unless a specific output frequency is required. A toroid inductor is desirable, but the author has successfully used the primary windings of a number of transformers, such as the Calrad CR40 and the Staco 023-1186.

the oscilloscope. Connect the oscilloscope ground to the circuit ground.

3. Connect 20 volts to the circuit with the positive lead on the top of the tank and the negative lead running to ground.

4. Observe the sine wave shape of the output.

5. If your oscilloscope is calibrated for frequency measurement, note the output frequency of the oscillator.

QUESTIONS

From the following list select a symbol or word to fill each of the blanks in the questions.

amount	inductance
amplitude	low
buffer	mutual
carrier	oscillator
electromagnetic	primary
envelope	rate
farad	resonant
frequency	secondary
harry	self
henry	tank
high	volt

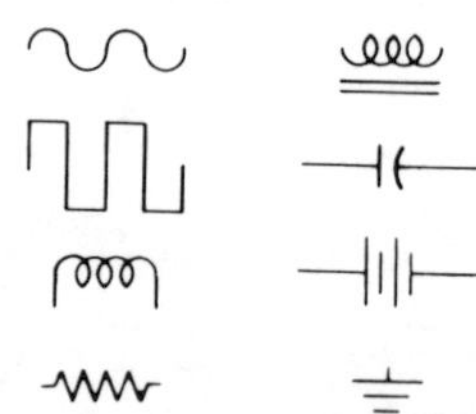

1. Just as a tiny guitar string can be lightly plucked and radiate a loud sound, whereas a heavy, long steel rod would radiate little, an antenna must be fed with a _________ frequency if much radiation is to occur.

2. The frequency necessary for radio transmission is produced by an _________.

3. Radiation from an antenna creates an _________ field.

4. The symbol for an air-core inductor is _________, but for an iron-core inductor it is _________.

5. The symbol for a capacitor is _________.

6. The unit of inductance is the _________.

   A RADIO TRANSMITTER

7. The unit of capacitance is the __________.

8. When a capacitor is placed in parallel with an inductor, a __________ circuit has been constructed.

9. A transformer consists of coils wound on an iron core. The winding used for input is called the __________ and others used for output are called __________.

10. The phenomenon that causes a voltage to occur in a transformer secondary when current changes in the primary is called __________ __________.

11. If a voltage is momentarily placed across the capacitor in the circuit described in question 8 and then removed, current will alternately flow clockwise and counterclockwise in the circuit. These alternations will be at the __________ frequency of the circuit.

12. The frequency considered in question 11 will **decay** in amplitude due to resistance in the circuit unless energy is added at the correct instant. This is accomplished by a device such as a Hartley __________.

13. To prevent variations in load on an oscillator that might affect its frequency, it is common design practice to feed the oscillator output into a __________ amplifier.

14. The frequency amplified by the buffer amplifier is called the __________ wave.

15. The outline of an amplitude-modulated wave is called its __________.

16. Lightning causes static because it makes changes in the __________ of the wave.

17. In amplitude modulation, the volume of the audio signal is represented in the modulated wave by the __________ the output is changed in __________.

18. In amplitude modulation, the __________ of the audio signal is represented in the modulated wave by the frequency at which the output is changed in __________.

19. In frequency modulation, the volume of the audio signal is represented in the modulated wave by the __________ the output is changed in __________.

20. In frequency modulation, the frequency of the audio signal is represented in the modulated wave by the __________ at which the output is changed in __________.

7

A Radio Receiver

It is the function of a radio receiver to select the proper signal from all of those broadcast by transmitters within range, to extract the original shape of the audio wave from the modulated wave, to amplify it, and to deliver it to a loudspeaker.

AM reception will be discussed first. For FM reception, see Section 7.9.

When you walk into a busy place, your ears are exposed to all kinds of noises making it difficult to separate one particular conversation from all the rest unless the speaker you want to hear is very loud or is right next to you. A radio antenna is in a similar situation. Signals from hundreds of transmitters are striking it with various intensities. Fortunately, it is possible to select just one of these and exclude the others by use of a **tuner.** The heart of this device is a tank circuit that can be adjusted to oscillate at the frequency of the carrier wave from the selected transmitter. When either the capacitor or the inductor is adjusted, a new station can be selected from all the other signals striking the antenna. In Figure 7.1 the

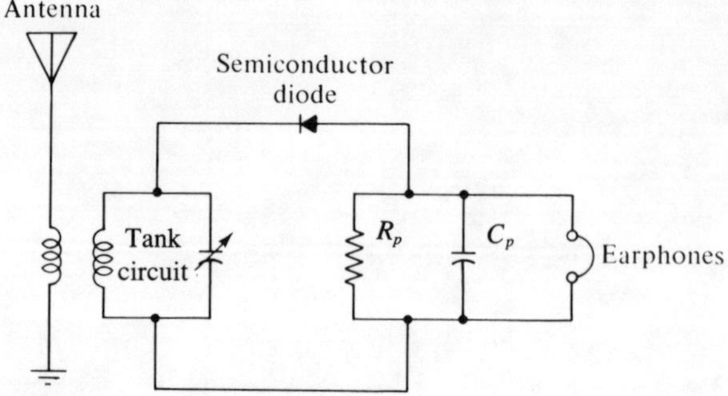

FIGURE 7.1   Radio receiver circuit

signal from the antenna is coupled to the tank circuit by means of a coil wound very close to the inductor coil of the tank circuit.

Signals from all sources cause alternating currents to flow between the antenna and ground. These induce voltages in the tank circuit inductor, but only one signal will cause the tank circuit to oscillate strongly. When the capacitor is adjusted to a value such that oscillation takes place in the tank circuit at the frequency of the carrier wave, currents can flow in the circuit illustrated. In other words, the tank circuit is tuned to the resonant frequency of the carrier wave.

7.3 *DETECTION*

In order to obtain the audio signal from the incoming wave that has been tuned in, a process called **detection** is employed. An incoming wave would have an appearance similar to that shown in Figure 7.2a. The audio wave

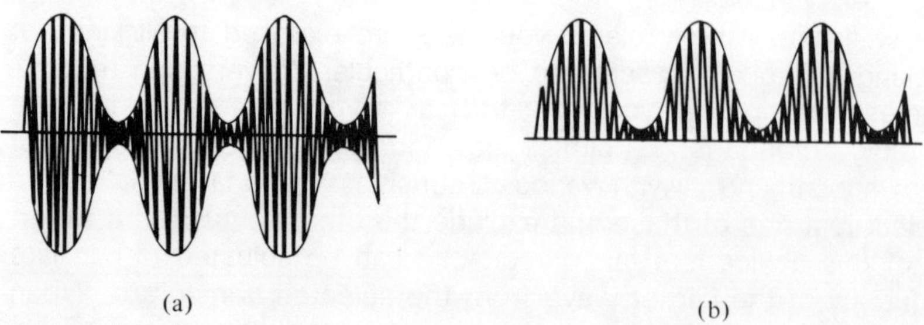

FIGURE 7.2   Rectifying of a carrier wave

shape is represented by the outline of either the top or the bottom portion of this carrier. However, when the top half is moving more positive, the bottom half is moving more negative. If these voltages appeared at the same time across a resistor they would nullify each other. What is needed, then, is a means of separating them. Only one is needed, so there is no real loss if one is completely clipped off as in Figure 7.2b. This separation is easily accomplished by a component called a **diode.**

DIODE　**7.4**

The name *diode* derives from the fact that there are two electrodes, a cathode and an anode. Either the semiconductor type described in Section 5.9, or the vacuum tube illustrated in Section 5.11, has the property of presenting a very high resistance to electron flow in one direction, but low resistance in the other.

There are many analogies to this process in other devices. Familiar to most people is the check valve in an automobile tire, which allows air to be pumped in but automatically closes to prevent air from leaking out.

In Section 5.9 it was pointed out that with a semiconductor junction of P-type and N-type material the resistance is low when a positive polarity is applied to the P-type material (the anode) and a negative polarity is placed on the N-type material (the cathode) (or the PN junction is forward biased). Applying the opposite polarity, or reverse bias, causes high resistance in the junction (see Figure 7.3).

An ideal diode would present zero opposition to a forward bias voltage and infinite resistance to reverse bias. Like most ideals, such perfection exists only in imagination. A small force called a **threshold voltage** is required in the forward direction to turn on the diode. For germanium, this threshold is about 0.2 volt, but for silicon it is approximately 0.6

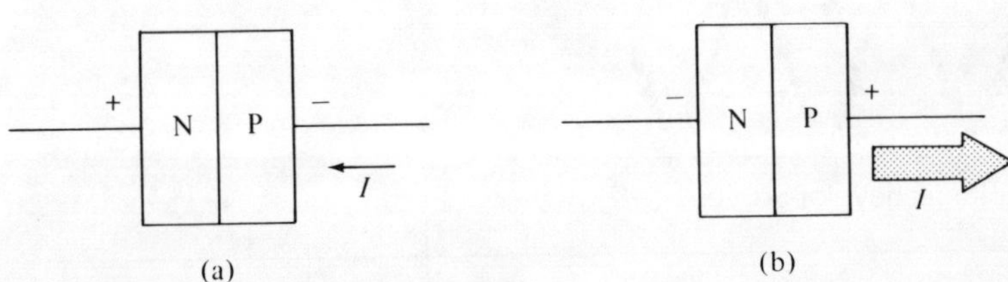

FIGURE 7.3　Semiconductor diode action

to 0.7 volt. In the reverse-bias mode there is **leakage** current. Usually leakage current is quite small, but it tends to be much greater in germanium than in silicon.

It is possible to raise the reverse-bias voltage high enough to overpower the barrier to current. However, such action usually destroys the diode, except with a specially designed type commonly called a **zener** diode. Zener diodes are normally operated only in reverse-bias mode to function as a voltage regulator. The point at which the diode begins to conduct when in reverse-bias mode is called the **breakdown voltage.*** The process of a diode passing only one-half of an AC wave is called **rectification.**

7.5 *AUDIO SIGNAL*

When the split-off (or **rectified**) wave of Figure 7.2b meets the capacitor and resistor (C and R) of Figure 7.1, the radio-frequency peaks and valleys are smoothed out because the capacitor opposes rapid changes in voltage and the resistor delays its discharges. The amount of capacitance is chosen so that it strongly opposes the very rapid RF changes, but "rides along" with the slower AF (audio) changes.

The resulting **filtered** wave resembles Figure 7.4. Note that the audio wave appears to be a sine wave, except that it has a jagged appearance. Actually, the illustration has been greatly exaggerated. The frequency of the jiggles of the line (RF) is far above what the human ear can hear (and more than the mass of the diaphragm of the earphones can follow). As a result, the signal heard in the earphones is as shown in Figure 7.5.

The whole process of cutting the carrier wave in half (**rectifying**) and of smoothing out the shape of the audio wave is called **de-**

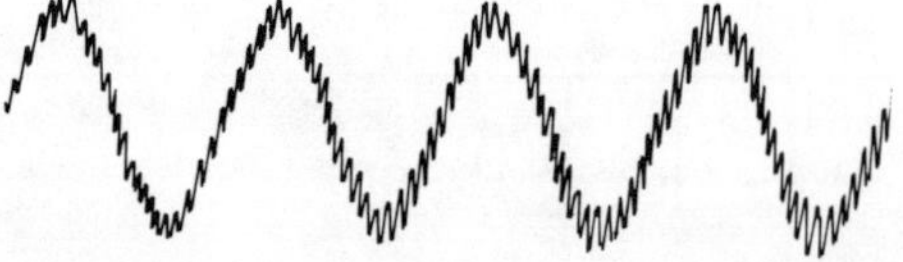

FIGURE 7.4   Voltage developed across R_p

---

*Terms such as **zener** voltage or **avalanch** voltage are sometimes used, but these expressions more properly identify critical points in the breakdown process.

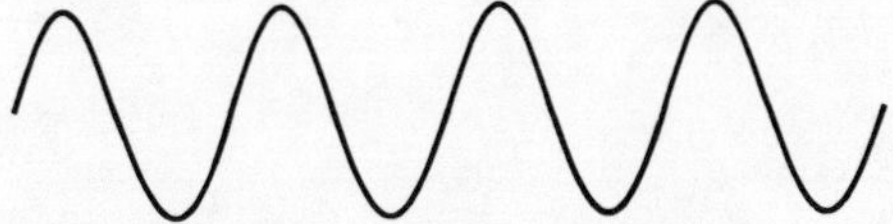

FIGURE 7.5 Detected audio wave

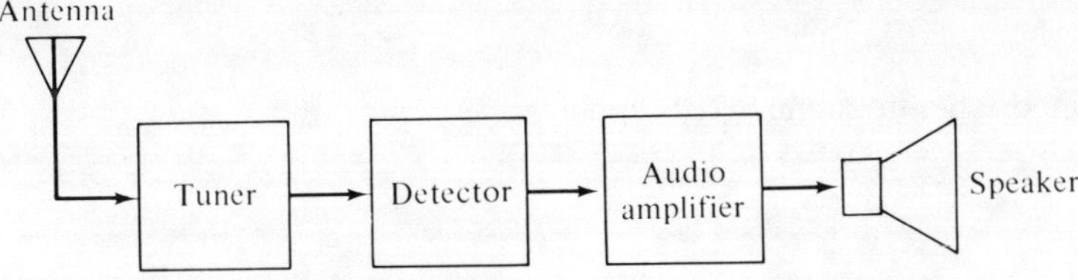

FIGURE 7.6 Radio receiver

tection. The detected audio wave can be amplified and fed to a loudspeaker in much the same way as the signal from a microphone was processed in the public address system previously described. In block diagram form a simple receiver appears as shown in Figure 7.6.

SUPERHETERODYNE **7.6**

In the early days of radio, simple circuits such as the one just described were commonplace. However, such receivers are not very **sensitive** nor **selective.** That is, they cannot amplify a very weak signal enough to be practical (**sensitivity**), and they cannot clearly separate signals that are close together in frequency (**selectivity**). Furthermore, if several stages of radio-frequency amplification are employed, each stage must be tuned to the carrier frequency. The result is that whenever a different station is tuned in, every stage of RF amplification must be adjusted to the frequency of the desired station. The block diagram in Figure 7.7 illustrates this type of receiver circuit. For most uses such a complex tuning procedure is inconvenient.

A great improvement was made in 1918 when Major Armstrong invented a circuit called the **superheterodyne.** Knowing the origin of this word will help in understanding its meaning.

Heteros is a Greek word meaning *different.* Similarly, *dyne* means *force.* Literally, *heterodyne* means *different forces.* As it turns out, when a wave is modulated by another it is equivalent to the adding and

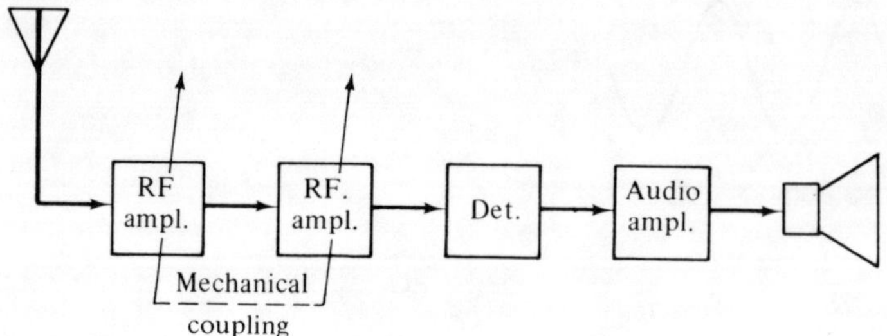

FIGURE 7.7 Tuned radio-frequency (TRF) receiver

subtracting of the different forces. *Heterodyne* is a synonym, therefore, for *modulate*. *Super* is a Latin-derived prefix meaning *above* or *over*. If a receiver is tuned to a carrier wave that has already been **heterodyned** (modulated) at the transmitter, and then the receiver proceeds to modulate that carrier, with yet another frequency, the original carrier wave has been superheterodyned.

7.7 SIDEBANDS

In Chapter 6 **mixing** was exhibited when an audio frequency was used to modulate (or heterodyned with) a carrier-wave frequency. The envelope or outline of the resulting wave form showed the frequency of the audio wave (see Figure 6.10). Such a development was relatively obvious. However, something else also happens when frequencies are mixed that is subtle and seems more mysterious. When two frequencies are heterodyned, two new frequencies are created. One of these will be a frequency that is equal to the sum of the two originals, and the other frequency will be equal to their difference. For instance, if a 600-kilohertz carrier wave is heterodyned with a 3-kilohertz audio wave, new frequencies of 603 kilohertz and 597 kilohertz will result. Since these new frequencies appear on "either side" of the carrier frequency, they are called **sidebands.**

7.8 LOCAL OSCILLATOR

Superheterodyne receivers contain a circuit called the **local oscillator.** A variable element in the tank circuit of the local oscillator is mechanically coupled to the receiver's tuner that selects the frequency of the desired station. The tank circuit of the local oscillator is so designed that it will oscillate at a frequency that is always a fixed amount higher or lower than the carrier frequency selected.

The output of the local oscillator is then combined with the received carrier wave in a circuit called a **mixer** where new sidebands or frequencies are created. One sideband will be a frequency that is the sum of the carrier frequency and the local oscillator frequency. The other will be a frequency that is equal to their difference (called the **difference frequency**). If the mixed waves are coupled to a tank circuit tuned to the difference frequency, the tank circuit will **reject** the carrier frequency and the sum frequency but will oscillate at the difference frequency. The difference frequency will also be modulated by the original audio waves that modulated the carrier wave.

The difference frequency most often used in standard broadcast radios is 455 kilohertz. It is called the **intermediate frequency (IF)** to distinguish it from the broadcast radio frequency (RF).

Since the intermediate frequency is always the same within each individual radio (as established by the difference in the sizes of the mechanically-coupled tunable parts of the RF stage and the oscillator), no change in tuning is needed from one IF amplification stage to another when a different station is tuned in. These IF stages can, therefore, be designed to have the most ideal characteristics possible since they are not required to perform over a large range of frequencies.

DETECTOR   **7.9**

The output from the IF amplifier stages must be detected to recover the audio frequency and reject the IF. After this is done, the audio signal is fed to a power amplifier that powers a loudspeaker. The entire process of the superheterodyne is represented in the block diagram in Figure 7.8.

Suppose that a station broadcasting with a carrier frequency of 545 kilohertz is tuned in. As the tuning knob is turned to adjust the tank circuit of the RF amplifier to that frequency it also adjusts the tank circuit of the local oscillator so that it will produce a proportionately higher or lower frequency. In this example, assume that an intermediate frequency of 455 kilohertz is used and that the local oscillator frequency is higher than that of the carrier. The components of the tank circuit of the local oscillator are selected with values such that the oscillator frequency at this tuning will be 545 + 455 = 1000 kilohertz. This 1000-kilohertz signal will then be mixed with the received carrier-wave frequency

    A RADIO RECEIVER

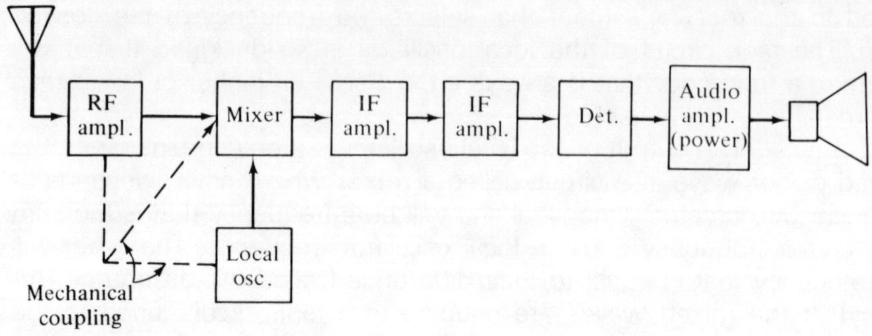

FIGURE 7.8 Superheterodyne

of 545 kilohertz. The result is audio-modulated sidebands of the sum and difference frequencies:

$$1000 + 545 = 1545 \text{ kHz}$$
$$1000 - 545 = 455 \text{ kHz}$$

Since the selected IF frequency is 455 kilohertz, the tank circuits of the IF amplifier are permanently tuned to select signals only at that frequency. Therefore, the carrier frequency with its sidebands and the 1545-kilohertz frequency with its sidebands are rejected. Only the IF frequency with its sidebands is accepted and greatly amplified. At the detector, the IF wave is rectified and smoothed so that only an audio wave remains. This wave is fed to the power amplifier that drives the loudspeaker.

7.10 *FM DETECTION*

Several methods are used for handling an FM wave after it has been tuned in so that the audio information can be extracted. Usually, the carrier is mixed with the signal from a local oscillator to superheterodyne so that succeeding stages can be operated at an intermediate frequency, just as in the AM receiver. (Since FM is generally operated at much higher frequencies than AM, the IF frequency employed is also much higher.) Following IF amplification, detection must be accomplished. One FM detector circuit is called a **discriminator.**

In Section 6.15, one reason for using FM was that it is more static-free than AM. This is due to the fact that most causes of static tend to change the amplitude of the carrier and create **spikes** on the wave envelope. Spikes can be removed before processing a wave through a discriminator because its proper operation depends only upon frequency changes, and amplitude changes can be eliminated.

The **clipping** of the wave can be done in an amplifier stage (called a **limiter**) where the amplifier is "overdriven." That is, the input signal is made large enough that the transistor or vacuum tube reaches its maximum current,* or is cut off, before the peak positive or negative voltages of the input signals are reached. As a result, all irregularities of amplitude are removed before the wave reaches the discriminator.

To understand the operation of the discriminator, consider the circuit in Figure 7.9. An alternating signal source coming to inductor L_1 will cause the tank circuit L_2C_1 to oscillate with it. Each oscillation will cause current to alternate also through the resistor. This AC current will cause point X to be alternately positive and negative with respect to Y. The closer the frequency of the source approaches the resonant frequency of the tank circuit, the greater the flow of current will be through the resistor.** Of course, the greater this current is, the greater will be the voltage swings of X with respect to Y.

Now a diode is placed in the circuit so that X can only go positive with respect to Y (see Figure 7.10).

Next, a similar circuit energized by the same source is put beside the first. In this circuit the resonant frequency of the new tank circuit (L_3C_2) is made lower than the resonant frequency of the original circuit containing L_2, C_1, and R_1. The source frequency is somewhere in between the two tank resonances (see Figure 7.11).

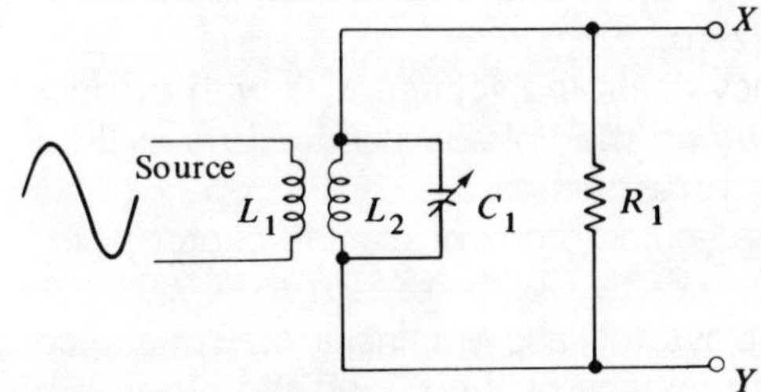

FIGURE 7.9 Effecting voltage change with frequency change

---

*This is termed **saturation.**
**The impedance of a tank circuit is maximum at resonance. At other frequencies it shunts (shorts) current around the resistor.

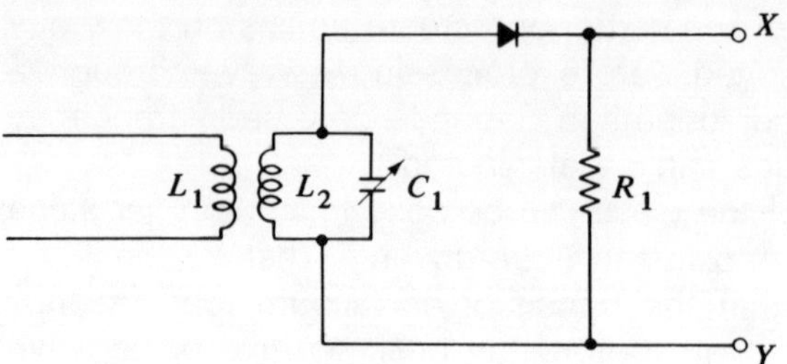

FIGURE 7.10   Prevention of negative charge at point X

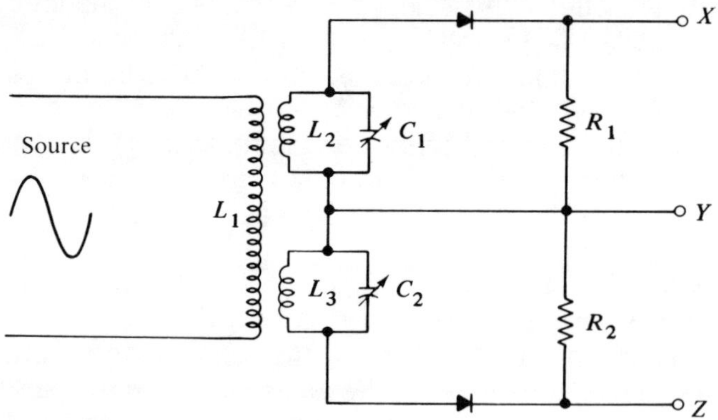

FIGURE 7.11   Basic discriminator circuit

When the source frequency is exactly centered between the resonant frequencies of L_2C_1 and L_3C_2, the voltage of X with respect to Y will be equal to the voltage between Z and Y. In other words, the difference between X and Z will be zero.

If the source frequency rises, the voltage of X with respect to Y will become greater than before and the voltage from Z to Y will be less. Therefore, X will be positive with respect to Z.

If the frequency of the source drops below the center point, the opposite happens. Point Z will be positive with respect to X.

The discriminator employs the above principles. The source is a tank circuit tuned to the center frequency of the IF and the other two tanks are tuned equally above and below this frequency. The FM modulated IF frequency that energizes the source tank will usually move back and forth between a frequency closer to the resonant frequency of L_2C_1 and one closer to L_3C_2. As it does so, an alternating voltage occurs between terminals X and Z that is at the frequency of the modulating audio signal. Thus, the modulation information has been retrieved as an alternating audio wave. Capacitors C_3 and C_4 have been added in the discrim-

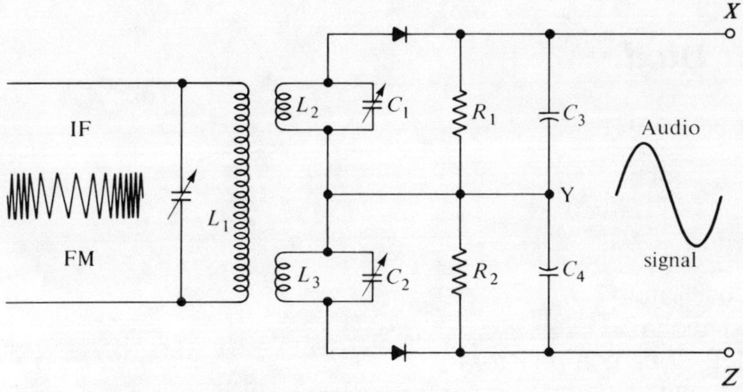

FIGURE 7.12 Discriminator circuit

inator circuit of Figure 7.12 to **filter** out any **spurious** RF signals that have come through the network.

AMPLIFICATION **7.11**

Following the detection of the audio signal from the FM carrier, all that remains to be done is to treat it like any other weak audio signal such as might come from a microphone pickup. Usually this means a stage or two of amplification and connection to a loudspeaker. The complete FM receiver is illustrated by Figure 7.13.

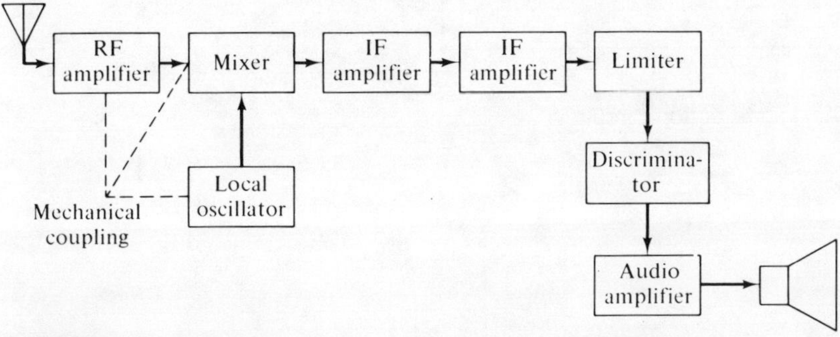

FIGURE 7.13 FM receiver

Action of a Diode

Equipment Needed

>   1—Audio oscillator
>   1—1N4001 diode
>   2—10-K 1/2-W carbon resistors
>   1—Oscilloscope
>   1—Breadboard

PROCEDURE

1. Construct the circuit illustrated in Figure A.
2. Turn on the oscilloscope and audio oscillator.
3. Set the oscillator on 1000 Hz.
4. Touch the oscilloscope probe to point *A* and observe the sine wave output of the oscillator. (Adjust the oscilloscope horizontal sweep rate until at least two full cycles are visible on the screen, but not so many as to obscure the sine wave.)
5. Touch the oscilloscope probe to point *B* and note that now only half of the sine wave is visible.

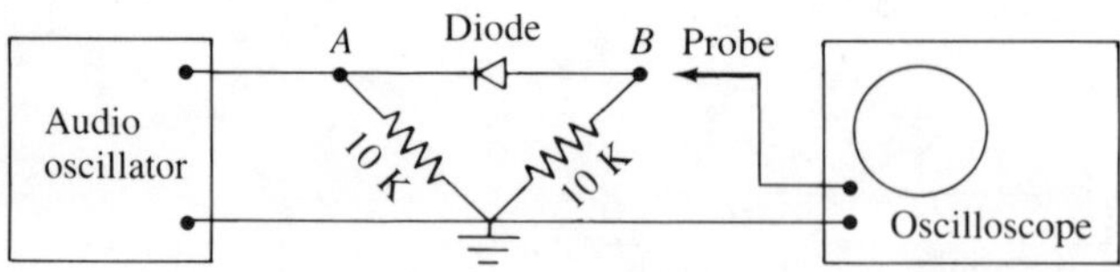

FIGURE A   Diode action circuit

QUESTIONS

From the following list select a word or number or combination thereof to fill each of the blanks in the questions.

audio	local	resultant	5
difference	modulate	sidebands	95
greater	one	static	100
intermediate	radio	superheterodyned	195
less	resonant	carrier	megahertz

1. A radio receiver selects one of many frequencies coming to its antenna by tuning a tank circuit so that its __________ frequency will be the frequency desired.

2. A diode is an electronic "check valve" that allows current in only __________ direction.

3. AM detection recovers the original modulating wave by half-wave rectification and smoothing of the __________ wave.

4. *Heterodyne* is another word for __________.

5. When a local oscillator is employed in a receiver the incoming signal is said to be __________.

6. In a receiver like the one in Question 5, most of the amplification is accomplished with the __________ frequency stages. This frequency is the __________ between the __________ frequency and the __________ oscillator frequency.

7. Limiting (clipping) of the signal received in an FM radio removes most of the effects of __________.

8. The greater the changes in frequency that are made from the center frequency of an FM station, the __________ the amplitude will be of the audio signal going to the audio amplifier of the receiver.

9. When a frequency is heterodyned, the result is two new frequencies called __________.

10. If an RF carrier frequency is 100 MHz and a local oscillator output is 95 MHz, the IF frequency is __________ __________.

8

The Oscilloscope

OSCILLOSCOPE USE 8.1

If the average electronic technician were asked what the most versatile piece of equipment is, the answer would probably be the oscilloscope. The oscilloscope can measure voltage, determine frequencies, and examine waveforms. Understanding the principle of its operation will simplify the examination of television operation in Chapter 9.

CATHODE-RAY TUBE 8.2

The heart of any oscilloscope is a **cathode-ray tube (CRT)**. This device is a type of vacuum tube that contains an **electron gun.** The electron gun emits electrons that are accelerated to a very high speed and focused into a beam. The beam of electrons is deflected in different directions until it strikes a glass surface (screen) coated with **fluorescent** salts. When the beam strikes the salts, it causes them to glow, appearing on the screen as a spot of light. The purpose of the CRT is to give the user an electronic "pencil" with which to write or draw pictures of otherwise invisible electronic signals or waveforms, most of which occur in a small fraction of a second.

8.3 *ELECTRON GUN*

The electron gun contains a cathode that is a metal cylinder (or can) enclosing a heating element. This cylinder is surrounded by a **control grid** that is a larger can with a hole in one end. (See Figure 8.1.) The heating element drives electrons from a special oxide coating on the surface of the cathode (a process called **thermionic emission**) to form a **space charge.** (See Section 5.12.) A negative charge applied to the outer cylinder (the control grid) controls the intensity (amount) of the electrons that can stream out through the hole and a large enough negative charge will cut off the electron flow entirely. The electrons flowing out of the hole tend to spread into an ever-widening beam. A very narrow beam is desired, however, and to achieve this a focusing device is used.

8.4 *FOCUSING*

While several methods of focusing may be employed, one method consists of two positively charged cylinders placed close together beyond the control grid. The second of these is larger than the other and has a higher

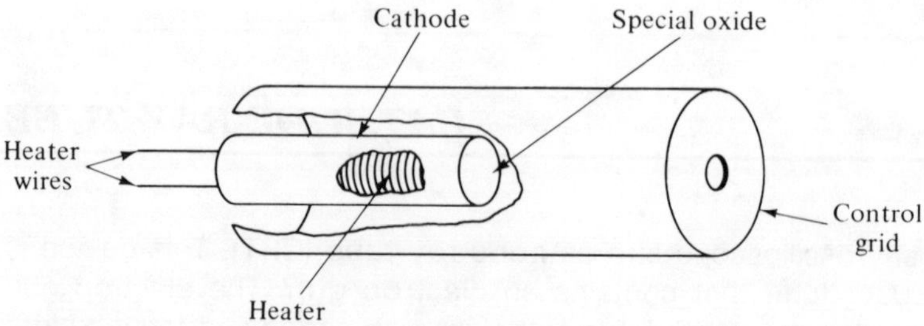

FIGURE 8.1 Cathode and grid assembly

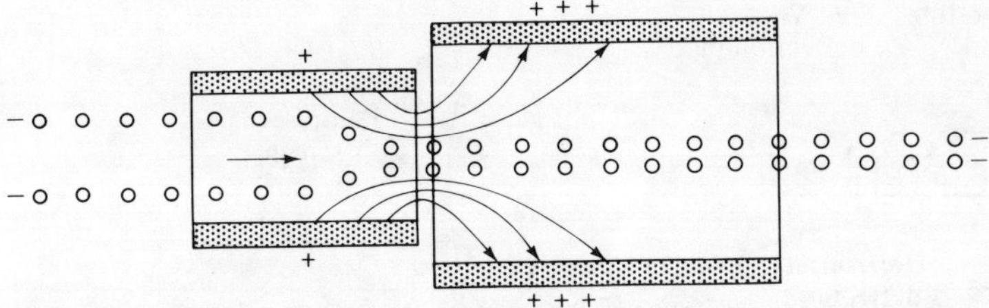

FIGURE 8.2   Focusing of electron beam

positive voltage. This condition creates an electric field between the two anodes that squeezes the high-velocity electrons into a narrow beam. As they attempt to follow the lines of force between the two anodes their speed is so increased that they cannot make the turn into the second very high-voltage anode and, thus, fly on through as a compressed stream. This effect is illustrated in Figure 8.2.

DIRECTING THE BEAM 8.5

The electrons passing through the focus and accelerating anodes travel at such high velocities that they fly right on past these positively charged surfaces without stopping. The high-velocity beam of electrons travels on until it strikes the fluorescent salts at the end of the tube. This would result only in a spot of light at the center of the fluorescent screen if it were not for the electrodes that are employed to bend the beam. These electrodes consist of two pairs of bent plates located just beyond the focusing anodes. One such pair is shown in Figure 8.3. If one of the plates is given a negative charge and the opposite one a positive charge, the beam will be bent in the direction of the positive plate. The negatively charged electrons are attracted toward the positive electrode and repelled by the negative one. The velocity of the beam is too great to allow it to strike the **deflection plates,** but the beam will now make a spot of light at a different place on the screen.

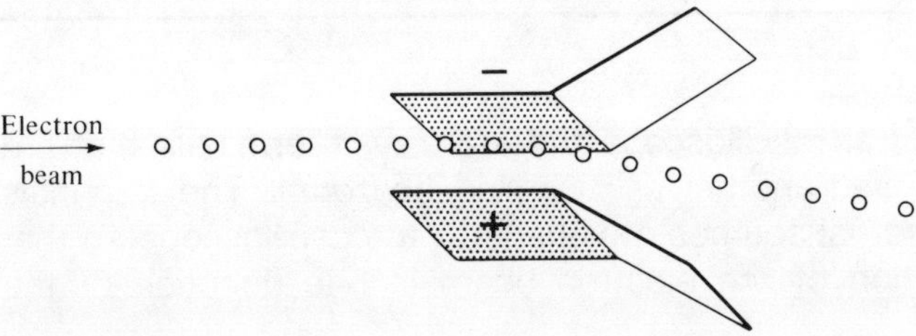

FIGURE 8.3   Deflection of electron beam

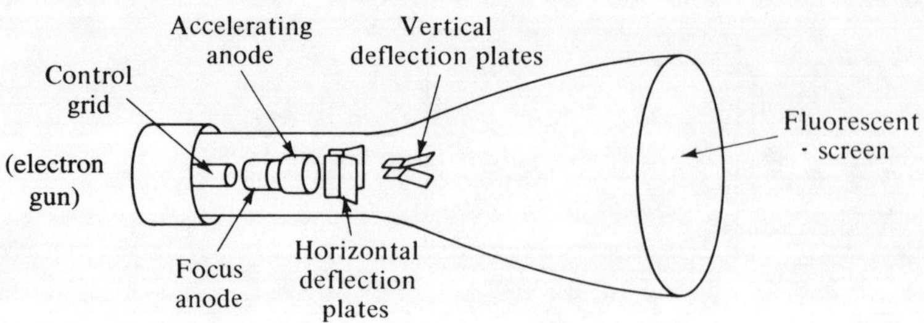

FIGURE 8.4 Cathode-ray tube

One pair of deflection plates is used to bend the beam from side to side and the other pair can bend it up or down. With the proper arrangement of charges on the plates the beam can be moved to cause the spot of light to appear anywhere on the face of the tube. The whole assembly of components is shown in Figure 8.4.

8.6 DEFLECTION CONTROLS

On the control panel of the oscilloscope are many control knobs for various purposes. Two of these are very elementary and are found on all oscilloscopes so as to position the impact point of the electron beam when no signals are being input. One of these adjusts the voltages on the horizontal deflection plates and the other performs the same function on the vertical plates. The **horizontal control** is turned counterclockwise to move the point to the left or clockwise to move it to the right. Likewise, a counterclockwise rotation of the **vertical control** will move the point up and a reverse rotation will move it down.

8.7 INPUT TERMINALS

On the control panel of an oscilloscope there are two terminals that are usually labelled *vertical* and another pair labelled *horizontal*. The terminals allow connection of external sources of voltage to the oscilloscope so that they will cause deflection of the electron beam in one direction or the other.

HORIZONTAL SWEEP **8.8**

Inside an oscilloscope is an electronic subsystem (including an oscillator) that can be switched (connected) to the horizontal plates to cause the beam to sweep across the screen from left to right at a constant speed. This circuit then shuts off the electron beam (with the control grid) while it rapidly reverses the polarity of the horizontal deflection plates to allow the sweep to begin again from left to right. Then the beam is turned on once more and the whole cycle is repeated. This subsystem is called a **horizontal sweep circuit** and can be adjusted to sweep the beam (**trace**) with a wide range of speeds. The action taken to shut off the electron beam during the **retrace** is called **return-trace blanking.**

The face of the CRT usually has a grid of horizontal and vertical lines called a **reticle.** Some oscilloscopes have a dial by which the horizontal sweep can be adjusted to discrete speeds so that each space on the reticle grid will measure a time interval. This allows an operator to display a cycle or more on the screen and measure accurately the period. The reciprocal of this value is then calculated to obtain the frequency. Less sophisticated oscilloscopes have an adjustment for sweep frequency, but it is not calibrated.

WAVEFORMS **8.9**

The fluorescent salts used on the CRT face have a quality called **persistence** so that they glow for a while after the electron beam is gone. For this reason a moving beam appears as a line if the movement is fast enough.

The sweep circuit makes it possible to view on the face of the cathode-ray tube the graph of a waveform. If the signal to be viewed is connected to the vertical deflection plates it will cause the beam to be moved up when the voltage is positive and down when it is negative. If the input voltage is changing rapidly as with a sine wave, and the sweep circuit is not operating, a bright vertical line will appear on the tube face. If the sweep circuit is now turned on and adjusted so that the beam is swept across the tube during the time that it takes for the signal to go from zero-to-positive-to-negative-and-back-to-zero, the waveform of one cycle will be displayed. Figure 8.5 illustrates several conditions that will result from connecting a 60-Hertz line voltage to the vertical input terminals.

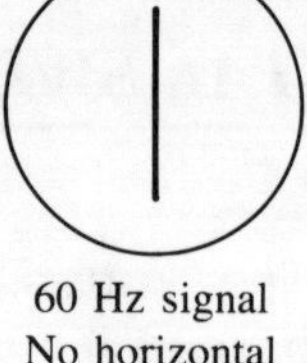
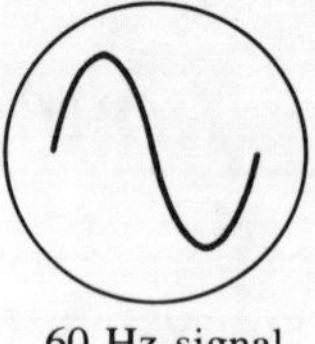
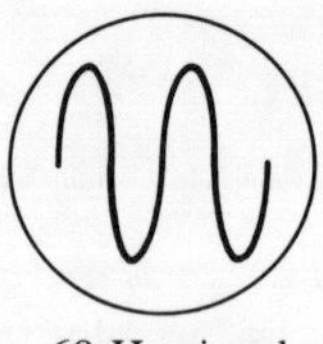
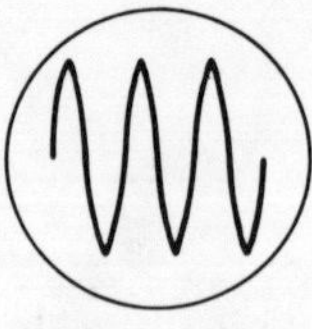

60 Hz signal
No horizontal
sweep
(a)

60 Hz signal
60 Hz horizontal
sweep
(b)

60 Hz signal
30 Hz horizontal
sweep
(c)

60 Hz signal
20 Hz horizontal
sweep
(d)

FIGURE 8.5   Effect of horizontal sweep frequency on sine wave display

In Figure 8.5a there is no horizontal sweep at all so that the variations in amplitude of the signal applied to the vertical terminals simply move the spot on the CRT up and down. The persistence of the phosphors makes this appear as a vertical line.

Figure 8.5b appears as a single cycle of a sine wave because the sweep moves across the tube in the time it takes for the voltage of the input signal to rise to its peak, fall to the negative peak, and return to zero. The return-trace blanking has cut off the electron beam at the end of the cycle so that the return sweep does not show.

Figure 8.5c shows two cycles displayed, because the input signal has time to go through two cycles while each horizontal sweep is being made.

Therefore, it should be obvious that a sweep frequency that is one-third the frequency of the input signal will produce three cycles for display, as illustrated by Figure 8.5d. In other words, the number of cycles displayed will be the input signal frequency divided by the sweep frequency.

8.10 *VERTICAL AMPLIFIER*

The signals an oscilloscope is usually used to examine are nearly always too weak to be applied directly to the deflection plates. For this reason, a **vertical amplifier** is provided to make it possible to move the beam drastically under the control of a very tiny signal. A control knob is provided to adjust the amount of amplification, and it can be calibrated so that the amount of vertical deflection of the spot on the screen can measure an input voltage.

VERTICAL AMPLIFICATION OR ATTENUATION 8.11

The best oscilloscopes provide for a calibrated amplification or **attenuation** (reduction of amplitude) of the signal fed to the vertical input. This means that a dial can be set that will make each vertical division of the reticle measure some fraction or multiple of a volt. (Other instruments may merely have a resistor network that permits less precise attenuation of the input in steps, usually in multiples of 10.)

Consider the following examples. With the horizontal sweep turned off, suppose a 2-volt **peak-to-peak*** signal applied to the vertical input causes the line in Figure 8.5a to be 5.0 centimeters long when the **step-attenuator** knob is set at *X1*. Now replace the 2-volt input with an unknown voltage, assuming for this example that the new voltage causes the line to become so long that the ends disappear off the edge of the screen. Turning the step-attenuator knob makes the line shrink until the ends can again be seen. If attenuation is increased until the line is again 5.0 centimeters long, and the step-attenuator knob is set at *X100*, the input voltage must be 100 times as large or 200 volts. If the line is 6 centimeters long with the *X100* setting, the unknown voltage is 600/5 of 2 volts, or 240 volts. If it is 8 centimeters, the unknown voltage is 800/5 times 2 volts, or 320 volts (and so on).

The oscilloscope is an excellent device for measuring alternating voltages since it can be used for a wide range of frequencies. Most voltmeters are accurate over a very narrow frequency range. In addition, the oscilloscope displays a signal's waveshape. This information is often very important and is beyond the capability of a voltmeter.

QUESTIONS

From the following list select a word, number, or combination thereof to fill each of the blanks in the questions.

attenuated	grid	space	2
bias	horizontal	speed	4
blanking	negative	sweep	40
charge(s)	plates	thermionic	57
circuit	reticle	trace	400
control	return	vertical	2800
glow	small	voltage	3200

---

*Peak-to-peak voltages are measures of the voltage changes from the maximum positive voltage reached to the maximum negative voltage attained in each cycle.

1. Electrons are driven off the cathode of the CRT by __________ emission.

2. The cloud of electrons that forms around a heated cathode is called __________ __________.

3. The flow of electrons from the cathode toward the anodes is restricted and varied by a __________ voltage applied to the __________ __________.

4. The electrons flowing toward the focusing anode pass by it because their __________ is too great.

5. In an oscilloscope, the focused beam is bent by opposite __________ applied to deflection __________.

6. When an electron beam strikes phosphors on the CRT face the phosphors __________.

7. Persistence is the characteristic of the phosphors that causes them to __________ some time after the beam has been removed.

8. The device within an oscilloscope that can be adjusted to sweep the beam from left to right at a constant speed is called the __________ __________ __________.

9. After the device described in question 7 has moved the beam across the tube, a negative voltage is applied to the control grid to provide __________ __________ __________, so that the return sweep does not show on the tube face.

10. If the sweep frequency is 400 Hertz, and 7 cycles appear on the tube face, the vertical input frequency is __________ Hertz.

11. Horizontal and vertical input amplifiers are usually provided on oscilloscopes because input signals are frequently very __________.

12. If one volt peak-to-peak causes a 1-inch vertical trace on an oscilloscope when the vertical step-attenuator is on *X1*, then the vertical input is __________ volts peak-to-peak if the trace is 0.4 inch long with the attenuator set at *X100*.

13. Focusing of the oscilloscope is obtained by adjusting the __________ between the accelerating and focusing anodes.

14. A grid of horizontal and vertical lines called a __________ is usually in place on the face of the oscilloscope CRT.

15. When the amplitude of a signal is reduced, it is said to be __________.

9

Television

TWO MODULATION SYSTEMS 9.1

Computers and robotics have captured the excitement and headlines of today's news, but television (TV) remains a vital element of our lives. In its entirety, television appears to be very complicated, but it is just a system composed of a number of simpler subsystems. In fact, a television set can be thought of as an AM receiver, an FM receiver, and an oscilloscope all combined into one unit.

PICTURE TRANSMISSION 9.2

The basic principle for the transmission of a picture is the same as for the transmission of sound in a radio system. Of course, a different transducer must be used to produce the information to be carried, but as with the previously described audio system, the signal produced by a transducer (camera) is amplified, and then used to amplitude modulate a carrier wave produced by an oscillator. The resulting RF wave is then radiated from a broadcast antenna to be received by individual television sets. The receiver superheterodynes the received signal, amplifies it with IF stages, detects it, and then uses the result to control the output of a cathode-ray tube similar to that used in an oscilloscope.

9.3 *TV AUDIO TRANSMISSION*

While the picture is transmitted by amplitude modulation, the sound for television is handled by frequency modulation (FM). Of necessity, TV transmission must be done with a **very-high-frequency** (**VHF**) or **ultrahigh-frequency** (**UHF**) carrier wave because the modulation required creates a wide spectrum of sidebands. There would not be room for one TV station in the entire standard AM broadcast band, much less all of the TV stations available. Since these high frequencies must be used, the range for the picture signal will be short. Frequencies above approximately 40 megahertz travel in a straight line as does light. Since they do not follow the curvature of the earth they radiate into space, and (depending on the antenna height and intervening obstacles) are not useful beyond about 50 miles. There is no reason to attempt to transmit the sound any farther than the picture. The use of FM transmission for sound has the advantage that it eliminates noise caused by natural disturbances (static), because these affect mainly the amplitude (and not the frequency) of the transmitted wave. The use of higher frequencies than have been common in the AM broadcast band has resulted in the transmission of a broader band of audio frequencies with the consequent result of **high fidelity.**

9.4 *TV CAMERA*

Several different kinds of tubes are used in television cameras, but all types make use of a **scanning** technique. All camera tubes depend upon the **photoelectric effect.** Certain materials such as selenium and various semiconductor materials have the properties of either emitting electrons, or showing reduced resistance, when exposed to light. These qualities are known as the **photovoltaic effect** and the **photoconductive effect,** respectively. If a current is passed through a photoconductive device in series with a resistor, there will be a voltage drop across the resistor. If the light falling on the photoconductive device changes in intensity, its resistance will change, causing a varying voltage to appear across the fixed resistor. One common type of camera, called the **vidicon camera tube,** uses this principle.

All camera tubes use an optical lens (similar to a lens in a regular film camera) to focus an image of the scene to be transmitted on a surface on the face of the tube. The vidicon tube has two layers of deposits on the inner surface where the image is focused. First, a thin transparent coating of conductive material is deposited. Behind that, a coating of selenium is overlayed. The selenium is called the **photolayer.** As was mentioned before, selenium has the peculiar property that its resistance decreases when it has light striking it. At the other end of the tube's glass envelope is an electron gun similar to the one described in Section 8.3. The image to be transmitted is focused through the optical lens and the tube's glass, and through the transparent conductive layer, onto the photoconductive selenium layer.

The electron beam from the gun is focused as was done in the oscilloscope tube and caused to sweep across the coated end of the tube in a succession of horizontal lines. The beam begins at the bottom left of the selenium layer (called a **target**) and proceeds to the right edge of the target. Then it retraces rapidly to the left edge (while being **blanked** as in the oscilloscope retrace) and begins another horizontal line. Each line is slightly higher than the last, and so after 240 horizontal sweeps have been "stacked," the beam reaches the top of the target. Here it is blanked again, and then rapidly swept back down to the bottom (called a *vertical retrace*) where the process repeats. All 240 lines is called a **field.**

When the beam strikes a point on the photolayer that has been brightly illuminated, most of the electrons flow through the lowered resistance into the conductive coating under it. Circuitry conducts this current through a load resistor producing a signal voltage (drop). When the beam strikes a poorly illuminated point in its sweep, the current conducted to the load resistor is small and the resulting signal voltage (drop) is also small. Thus a varying voltage is produced corresponding to light and dark areas on the selenium target. The rate of sweeping the image horizontally is 15,750 sweeps (lines) per second. This allows for 30 complete **frames** per second where each frame (480 horizontal lines) consists of two fields of the picture. A frame contains all the information in a complete television image. The first field covers all the odd-numbered lines of the frame from 1 through 479, and the second field covers the even numbers from 2 through 480. The time for an additional 45 lines is lost during vertical retracing.

If the image on a TV receiver screen is closely examined these horizontal lines are visible. In the interest of clarity, the number of

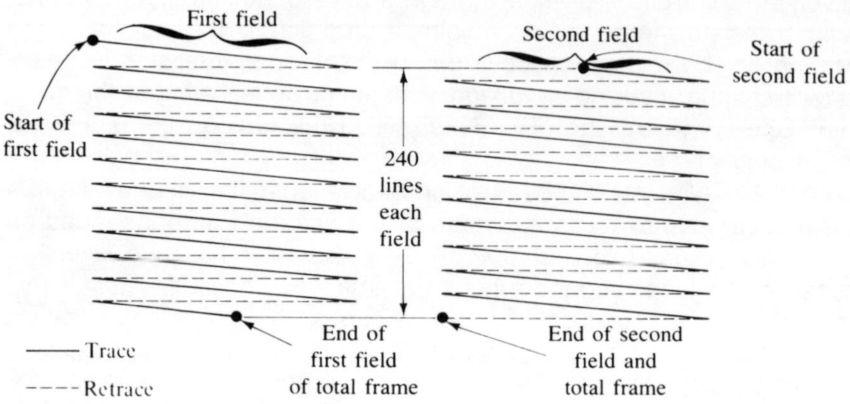

FIGURE 9.1   Diagram of complete frame showing two fields interlace

lines shown in Figure 9.1 has been greatly reduced. Remember that the lens that focuses the image also inverts it, so a camera tube starts its scanning at the bottom, while the receiver begins at the top. Notice that the lines (also called **traces**) of the two fields that make up a complete frame are **interlaced** so that the lines of the first field fall in between the lines of the second field. During the retrace time the sensing equipment is blanked by a much larger voltage pulse generated in the associated equipment. This pulse(s) permits **synchronization** of the transmitter and receiver.

9.5 *SYNCHRONIZING PULSE*

It would be useless to broadcast picture information if some method did not exist that would permit a receiver to keep in step with the camera at the transmitter. Indeed, few persons have missed the distress of a faulty receiver that would not stay "in sync." The picture seems to tear up into an unrecognizable series of light and dark streaks or, at best, to "roll over."

To achieve synchronization, oscillator circuits generate strong pulses that cause the camera to perform the retrace operations. A pulse is overlaid on the signal coming from the camera so that a receiver will cause its picture tube to go dark (blank) during the retrace and also perform its own retrace functions.

TV RECEPTION   9.6

The signals picked up by the receiver antenna are fairly complex. They contain the FM sound signal (**audio**) and the AM picture signal (**video**), including synchronization pulses. The sound carrier wave always has a frequency 4.5 megahertz higher than the video carrier. For instance, Channel 5 has a video carrier frequency of 77.25 megahertz and an audio carrier frequency of 81.75 megahertz. The receiver amplifies both together in an RF amplifier and then mixes (heterodynes) the amplified signals with a frequency generated by a local oscillator. The resulting IF frequency is amplified and detected before being directed to a subsystem called a 4.5-megahertz **trap.** At the trap, the 4.5-megahertz difference in audio and video signals allows the detected and amplified video signal to be split off and connected to the cathode-ray picture tube. The remaining FM signal is then detected to produce an audio signal that is amplified and connected to a speaker. The whole system is shown in Figure 9.2.

Of course, the operation of the video portion of the television receiver is not just the simple matter of connecting the detected and filtered AM signal. The synchronizing pulses received must cause an oscillator to apply voltages that will deflect the beam of the cathode-ray tube across the tube face (TV screen) in proper synchronization with the transmitter camera. The camera is transmitting video information as a varying voltage, and this must be applied to a grid in the receiver CRT so that the electron beam is controlled. If the video voltage is high, the beam-current flow is low, causing a relatively dark area on the receiver screen. If the received voltage is low, the corresponding area is bright. As each trace is completed, a high negative voltage must be applied to the grid so that no line will be produced during the retrace.

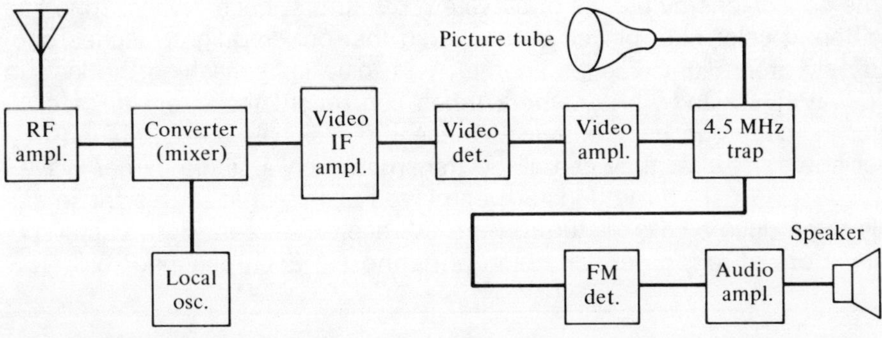

FIGURE 9.2 Intercarrier-sound IF system

The process is a lot like painting a picture with a spray gun that can be moved only horizontally except at the edge of the picture. Moving across the canvas, the trigger is pulled whenever a dark part of the picture is needed. When the end of the line is reached, the gun is swung back (retraced), dropped down two lines, and swept again to the right with the trigger pulled as required. If this process is repeated over and over for 240 sweeps to the right the picture will be roughly outlined. Moving again to the top of the picture (vertical retrace), line two can be sprayed and the process repeated, filling in all the even-numbered lines. This would complete the picture, and the process would start all over again with a new picture (frame).

In the early days of television the CRT used was identical to that in an oscilloscope, as described in Chapter 8. However, as television screens increased in size the beam could not be deflected enough using varying charges on deflection plates. Therefore, **magnetic deflection** was developed, using coils on the side of the CRT to move the beam. This type of deflecton works well in television where the sweep frequency is constant, but is impractical in oscilloscopes because the inductance of the electromagnetic coils creates different impedances at different frequencies.

9.7 *COLOR TELEVISION*

Even before black-and-white television was a commercial success, great effort was being made to transmit color. Several processes were devised, but the one finally adopted provides three video signals, one for each of the primary colors, red, green, and blue. Although the actual circuitry is quite complex, these three video signals are basically produced by three black-and-white camera tubes (like the vidicon) that all "look" through the same optical lens by means of a system of mirrors. Each camera tube has a different color filter placed before it, so that only red light reaches one, and only green light reaches another, with blue light reaching the last. In this way, black-and-white camera tubes can record the intensity of each color in the scene being viewed. These three signals are combined in a special way so that they can all be transmitted on a single carrier wave.

Two methods are employed at this time to use these signals to produce a color picture on the TV receiver screen. Both use a special pattern of red, green, and blue **phosphors.*** Each phosphor will glow

---

*These are the primary colors for transmitted light. The primary colors for pigments (paints) are usually considered to be red, yellow, and blue. The phosphors are fluorescent salts.

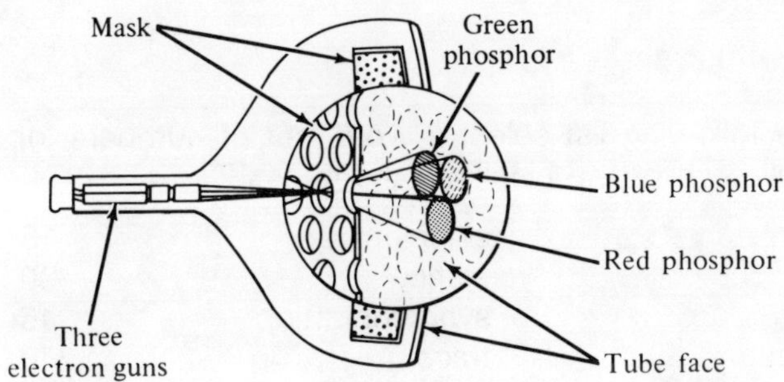

FIGURE 9.3   Activation of dot pattern by sweeping beams

only its characteristic color if struck by an electron beam. The oldest of the two methods arranges these phosphors into a dot pattern on the screen of the receiver picture tube. Just in back of this **phosphorescent** screen there is a perforated sheet called a **mask.** The phosphor dots are arranged in groups of three (one dot of each color), behind which for each group there is a single hole in the mask. Three electron guns are aimed at the mask, one for each of the three colors. The guns are positioned so that when all three guns are sending beams in the direction of one of the holes, the beams cross one another precisely at the hole. Beyond the hole in the mask the beam carrying the red signal is able to strike only the red phosphor; the green beam can strike only the green phosphor; and, the blue beam can strike only the blue phosphor. To help visualize this, see Figure 9.3 where a single dot pattern and mask hole are portrayed. Depending on the size of the screen, a picture tube might contain more than a million mask holes, and three times that many dots of phosphor.

Another method that has become popular places the phosphors in vertical stripes on the screen. The mask is an assembly of fine vertical wires, or contains vertical slots that perform a function similar to the mask with holes. The three electron beams come to a focus at the mask, but because of the angle at which each enters the openings in it, the red beam strikes only the red phosphors, the blue beam strikes only the blue phosphors, and the green beam strikes only the green phosphors.

In both methods, when a carrier wave modulated by color signals is received and detected, the appropriate intensities of the respective beams hit the phosphor pattern on the tube face and produce the illusion of many shades and hues of color. An equal intensity on all three phosphors appears as white.

QUESTIONS

From the following list select a word, set of numbers, or combination thereof to fill each of the blanks in the questions.

amplitude	gray	selenium	15
audio	green	static	30
black	grid	synchronization	45
blue	magnetic	trace	240
control	photoconductive	video	480
electric	photovoltaic	violet	525
frequency	red	white	15750

1. Transmission of the signals required for television is so demanding of bandwidth that very-high frequencies are required. __________ modulation is used for the video portion, and __________ modulation is used for the sound.

2. The TV camera must employ the same principle as photocells. The picture to be transmitted is swept a point at a time so that each point is sensed __________ times per second.

3. In the vidicon tube the photolayer is a coating of __________.

4. In a video camera pulses are generated that provide for the control of horizontal and vertical sweeps, as well as the blanking of each return trace and the repetition of fields and frames. These pulses are also transmitted to the receiver to provide __________.

5. In a TV receiver the __________ signal is detected first.

6. The horizontal sweep frequency is __________ sweeps per second.

7. Color TV provides for three beams to **converge** at openings in a mask just before striking phosphors on the tube face. When all three beams are of equal intensity, the result is a __________ spot.

8. The phosphors on the TV tube individually glow either __________, __________, or __________.

9. The electron beams in most television cathode ray tubes are deflected by a __________ field.

10. Photoelectric effects are of two types, __________ and __________.

11. The electron beam in a CRT can be turned off by applying a high negative voltage to the __________ __________.

12. The streak of light formed on the screen of the CRT by sweeping an electron beam across the tube is called a __________.

Part

II

Components and Theory

10

Power, Resistance, and RMS

Now that you have investigated some of the electronic devices that make our modern world pleasurable it should be apparent that one of the most frequently recurring phenomena in all the devices is a characteristic referred to as *impedance*. As discussed previously, this general term is applied to those functions that limit current. It can be inductive reactance, capacitive reactance, resistance, or a combination of any two or all three. In fact, all components exhibit all three types of impedance, but in some devices one or two of the types may be so small that they can be virtually ignored. For instance, a carbon resistor will display some inductive reactance and some capacitive effect with adjacent surfaces, but these impedances are usually ignored (except at ultrahigh frequencies) because they are insignificant compared to the resistance.

10.2 *RESISTANCE*

In previous chapters, resistance was described as roughly analogous to friction. When electrons are forced to move through a material, the structure of the atoms in the material determines the resistance to the flow. As the electrons move against this resistance, heat is generated. Heating is a unique characteristic of resistance that does not exist with other forms of impedance. The amount of heating is directly related to the current and voltage drop across the material.

10.3 *RESISTIVE CIRCUITS*

In nearly all electronic devices, resistance plays an important role, whether by design or because there is no way to avoid it. In most cases it is the same kind of problem presented by friction in mechanical design. Friction is a necessity in automobile design, for instance, because without it the tires would slip and slide all over the road. The brakes would not work, the fan belt would not drive the water pump or alternator, and countless other items would fail. On the other hand, if bearings, pistons, and gears could be free of friction, less power would be lost and parts would not wear out. In electronic design, resistors control current, provide necessary divisions of voltage, and radiate heat. Many circuits would not work without resistances. However, many devices would work better if resistance could be eliminated from them.

Because of this continued involvement of resistance in electronic devices, it is necessary to be able to calculate how currents and voltages are affected by resistances in various configurations.

10.4 *RESISTORS IN SERIES*

The calculations involving resistances in series were discussed in Section 3.5. The total resistance is simply the sum of the series resistances as indicated in the following formula:

$$R_T = R_1 + R_2 + R_3 + \ldots + R_n$$

RESISTORS IN PARALLEL 10.5

The calculation of resistances in parallel was discussed in Section 3.6. A formula was given, but its derivation was omitted. Now that we are examining resistance in greater detail, it is appropriate to explain where this formula came from.

In Figure 10.1 the electrons coming from the negative terminal of the battery travel to point X where they have several options (like cars coming to an intersection). In electronic circuits we call this point a **junction.** Some electrons will go "left" through R_1, some will flow "right," through R_3, and the remainder will flow "straight ahead" through R_2. The number of electrons (or cars) that enter an intersection must be equal to the number leaving it. This principle is known as **Kirchhoff's current law.** Stated formally, *the algebraic sum of currents at a junction equals zero.* In other words, if we consider the currents entering a junction as positive, and the currents leaving it as negative, the total of all currents will add to zero. Similarly, at point Y in Figure 10.1, the **branch** currents (through R_1, R_2 and R_3) enter the junction, and a current equal to their total must leave. All of the above may be stated in a formula as follows:

$$I_T = I_1 + I_2 + I_3$$

Each of the terms in the preceding equation has an equivalent value that can be calculated by Ohm's law:

$$I_T = \frac{E}{R_T} \qquad I_1 = \frac{E}{R_1} \qquad I_2 = \frac{E}{R_2} \qquad I_3 = \frac{E}{R_3}$$

If these equivalent terms are substituted into the previous formula, the result is the following:

$$\frac{E}{R_T} = \frac{E}{R_1} + \frac{E}{R_2} + \frac{E}{R_3}$$

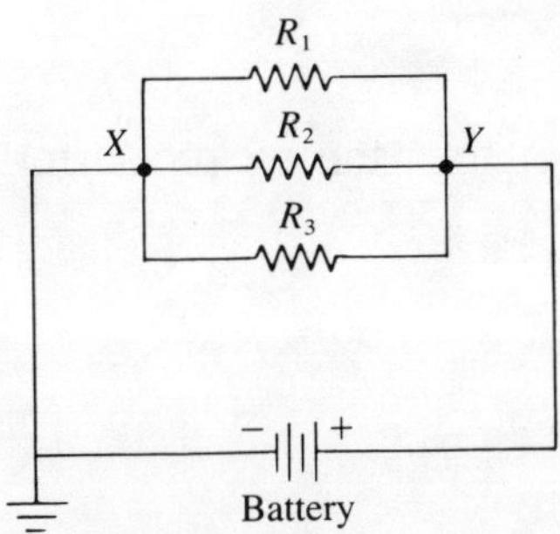

FIGURE 10.1 Parallel circuit

Since the quantity E is common to each term, we can divide through by E and the result is

$$\frac{1}{R_T} = \frac{1}{R_1} + \frac{1}{R_2} + \frac{1}{R_3}$$

We want R_T rather than its reciprocal, so we find the reciprocal of both sides of the formula and the result is the formula given in Chapter 3.

$$R_T = \cfrac{1}{\dfrac{1}{R_1} + \dfrac{1}{R_2} + \dfrac{1}{R_3} + \ldots + \dfrac{1}{R_n}}$$

When only two resistors are in parallel, the formula can be simplified to the following:

$$R_T = \frac{R_1 R_2}{R_1 + R_2}$$

This equation is commonly referred to as the *product-over-sum* formula. It is convenient to use and should be memorized.

Consider a circuit with a 60-ohm resistor in parallel with one of 40 ohms. With the product-over-sum formula,

$$R_T = \frac{60 \times 40}{60 + 40} = \frac{2400}{100} = 24 \text{ ohms}$$

Even if there were additional resistors in parallel with the first two, the formula could be applied. If one of 16 ohms and one of 4 ohms were in parallel with the 60-ohm and 40-ohm resistors, then

$$\frac{16 \times 4}{16 + 4} = \frac{64}{20} = 3.2 \text{ ohms}$$

and

$$\frac{24 \times 3.2}{24 + 3.2} = \frac{76.8}{27.2} = 2.82 \text{ ohms}$$

This method is sometimes easier than the reciprocal formula which would yield the same result.

$$\frac{1}{60} + \frac{1}{40} + \frac{1}{16} + \frac{1}{4} = .0167 + .0250 + .0625 + .2500 = .3542$$

$$\frac{1}{.3542} = 2.82 \text{ ohms}$$

CURRENT DIVISION IN A PARALLEL PAIR 10.6

Two resistances in parallel is such a common configuration in electronics that it is frequently helpful for a technician to have a formula that will determine quickly how a given current will divide between the two branches. As we derive this formula, please refer to Figure 10.2.

According to the formula, the total resistance of the two parallel resistors in Figure 10.2 is

$$\frac{R_1 R_2}{R_1 + R_2}$$

Ohm's law tells us that the voltage across the pair must be the total current I_T multiplied by this resistance. Thus,

$$E_T = \frac{I_T R_1 R_2}{R_1 + R_2}$$

The current that will flow through R_1 must be the total voltage E_T divided by R_1. Therefore,

$$I_1 = \frac{I_T R_1 R_2}{R_1 (R_1 + R_2)} = \frac{I_T R_2}{(R_1 + R_2)}$$

This result can be stated as follows: *When current divides in a parallel pair of resistors, the current in each resistor is the product of the total current and the other resistor divided by the sum of the two resistances.*

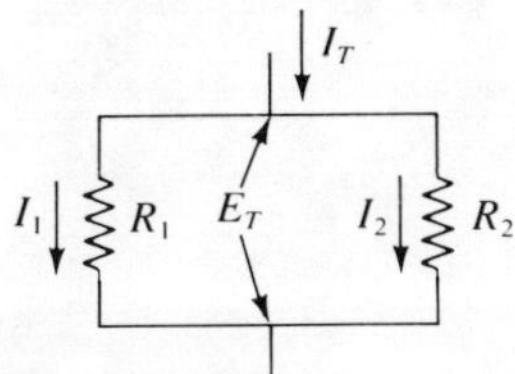

FIGURE 10.2 Current division

10.7 *SERIES-PARALLEL CIRCUIT*

Of course, a great many circuits in electronics are series-parallel hookups where some resistances are in series and some are in parallel. A typical transistor circuit (shown in Figure 10.3) illustrates such a condition.

To comprehend fully that series and parallel connections are demonstrated in Figure 10.3, recall that the transistor is itself a type of resistor. To calculate the resistors required for the proper operation of such a circuit it is easier to consider first the principles involved in some very simple circuits.

Refer to the circuit shown in Figure 10.4. If we wish to find the total current in such a circuit the procedure is a series of easy calculations. First, find the total resistance of the innermost section. Adding 65 ohms to 15 ohms gives a result of 80 ohms. Then sketch or imagine the simplified circuit, as in Figure 10.5. Find the resistance of the parallel section.

$$R = \frac{20 \times 80}{20 + 80} = \frac{1600}{100} = 16 \ \Omega$$

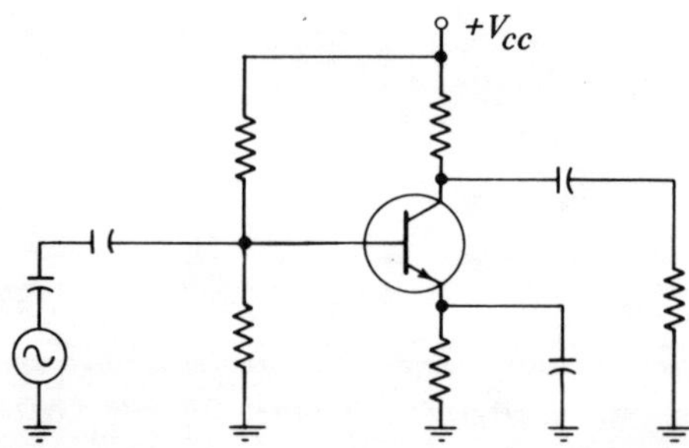

FIGURE 10.3   Typical amplifier with resistors in series-parallel connection

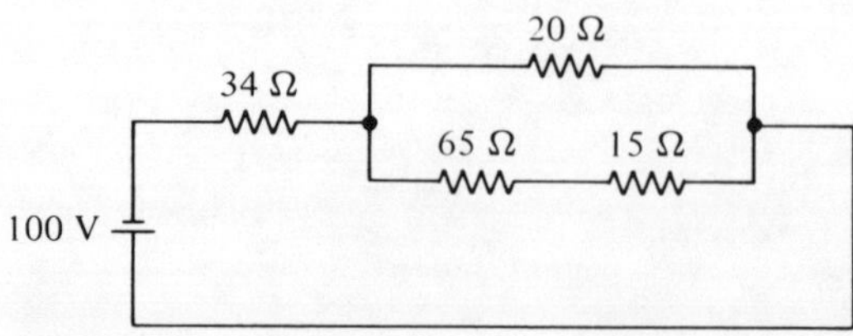

FIGURE 10.4   Example of series-parallel connection

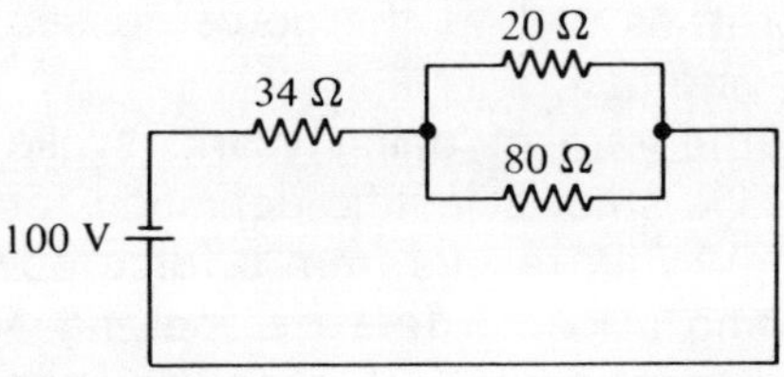

FIGURE 10.5 Simplified circuit of resistors

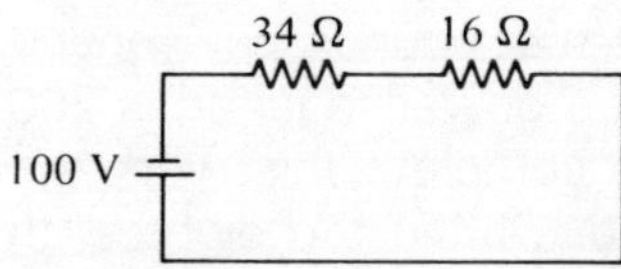

FIGURE 10.6 Simplified circuit of resistors

Sketch this further simplified circuit (Figure 10.6). Find the total resistance by adding the resulting two resistances in series.

$$R_T = 34 + 16 = 50 \ \Omega$$

Finally, find the total current.

$$I_T = \frac{E_T}{R_T} = \frac{100}{50} = 2 \ \text{A}$$

POWER **10.8**

In Chapter 1 electronics was contrasted with electricity as a control function rather than a power function. To illustrate, a piezoelectric crystal is primarily an electronic device, and an electromagnet or a motor is primarily an electrical component. The crystal provides a control signal at very low power whereas the electromagnet or motor exist primarily to deliver mechanical power. In almost all cases in electronics, however, there is a need for power output after the control function has been achieved. As an example, the last stage of a radio receiver usually is an audio (or power) amplifier that provides the necessary drive for the electromagnet in a loudspeaker. Considerable work must be done to create the compressions of air we recognize as sound. It is necessary to understand power, therefore, and be able to calculate requirements for it, as well as the power losses that are inevitable in any system or subsystem.

Power is, by definition, the *rate* of doing **work.** While everyone knows what "work" is, there is a strict scientific definition for *work* that is not so universally understood. Work results from a force operating through a distance. That is, pushing a load 5 feet by applying a force of 10 pounds through the entire distance results in 50 **foot-pounds** of work (5 × 10 = 50). Likewise, by lifting a 25-pound weight a distance of 4 feet, 100 foot-pounds (4 × 25 = 100) of work is done.

Power is the *rate* of doing work, so lifting the 25-pound weight through 4 feet in 1 second develops 100 foot-pounds per second of power. If the job is done in one-half second, the power required is 200 foot-pounds per second. If 2 seconds is used to complete this work the power needed is only 50 foot-pounds per second. During the eighteenth century, James Watt (the inventor of the modern steam engine) made measurements of the rate at which horses could do work so that he could relate the power of his engines to something his prospective customers would understand. From his tests he established 550 foot-pounds per second as one **horsepower,** thus creating one common unit of measurement.

10.9 *POWER DEVELOPED IN RESISTANCE*

As was mentioned in Section 10.1, power in the form of heat is developed in a resistance when current flows through it. Sometimes this is a desirable thing as with the element of a heater or an incandescent light bulb. At other times the heat developed is a loss. In power-transmission lines, for instance, the power lost as heat due to the resistance of the lines can be very costly. Many times the heat produced due to power loss in electronic circuits creates cooling problems. Excess heat can cause malfunction or destruction of many components and is particularly critical when semiconductors are used.

The common electrical unit for power is the **watt,** named after James Watt who established the horsepower unit. The watt is a much smaller unit, however, there being 746 watts per horsepower.

Electrical power results from current and voltage operating together. Power can be expressed as the product of the two. In a steady-state DC circuit, this is expressed as

$$P = EI$$

where

$$P = \text{power (in watts)}$$

$$E = \text{EMF (in volts)}$$

$$I = \text{current (in amperes)}$$

In Chapter 3, Ohm's law showed that $I = E/R$. If the expression E/R is substituted for the I in the power formula the result is a new formula for power:

$$P = EI = E \times \frac{E}{R} = \frac{E^2}{R}$$

Also, from the Ohm's law formula showing that $E = IR$, the replacement of the E in the original power formula results in a third power formula:

$$P = EI = IR \times I = I^2R$$

One aspect that frequently perplexes students is the effect of adding (or subtracting) resistors in series or parallel circuits. Consider, for example, a circuit consisting of five resistors in series where each has a value of 10 ohms. If 100 volts is applied across the circuit the power dissipated by the resistors is calculated by the power formula as follows:

$$P = \frac{E^2}{R} = \frac{(100)^2}{10 + 10 + 10 + 10 + 10} = \frac{10000}{50} = 200 \text{ watts}$$

If the number of resistors is reduced to four,

$$P = \frac{E^2}{R} = \frac{(100)^2}{10 + 10 + 10 + 10} = \frac{10000}{40} = 250 \text{ watts}$$

The power **dissipated** is *increased* by reducing the resistance. If this seems difficult to comprehend, recognize that with the reduction in resistance, there is an increase in current. Power increases by the *square* of the current.

Now, consider the case of resistors in parallel. This time place five 10-ohm resistors in parallel across a 100-volt source. The voltage across each one will be 100 volts so that the power in each is

$$P = \frac{E^2}{R} = \frac{(100)^2}{10} = \frac{10000}{10} = 1000 \text{ watts}$$

Therefore, the total power dissipated by five resistors is 5000 watts. For only four resistors it would be only 4000 watts. It is obvious, then, that reducing the number of parallel resistances in a circuit will reduce the power that is dissipated. However, if any of the resistors are merely reduced in value (rather than being removed) the power used will *increase*. For instance, replace one of five 10-ohm resistors with a 2-ohm resistor and the power consumed by it will be

$$P = \frac{E^2}{R} = \frac{(100)^2}{2} = \frac{10000}{2} = 5000 \text{ watts}$$

The total for all five in parallel would then be 9000 watts.

10.10 *POWER IN AC CIRCUITS*

A great many electronic circuits employ alternating rather than direct current. Power is developed in resistors by the alternating current, but from instant to instant the amount varies because the current is constantly changing with the changing voltage.

In Section 10.9 it was shown that power developed in a resistor can be calculated by the formulas

$$P = I^2R \quad \text{or} \quad P = \frac{E^2}{R}$$

These same relationships apply with alternating current but are expressed in terms of **instantaneous** values. To make it clear that the values are instantaneous, lowercase letters are used in the formulas as follows:

$$p = i^2R \quad \text{or} \quad p = \frac{e^2}{R}$$

Figure 10.7 illustrates various instantaneous values of current or voltage that occur with a sine wave. The value of amplitude at instant *a* is greater than zero but less than the maximum, which occurs at instant *b*. At instant *c* the amplitude is zero. By instant *d* it is again maximum but with opposite polarity and direction. It declines again through *e* to reach zero again at *f*, and so on.

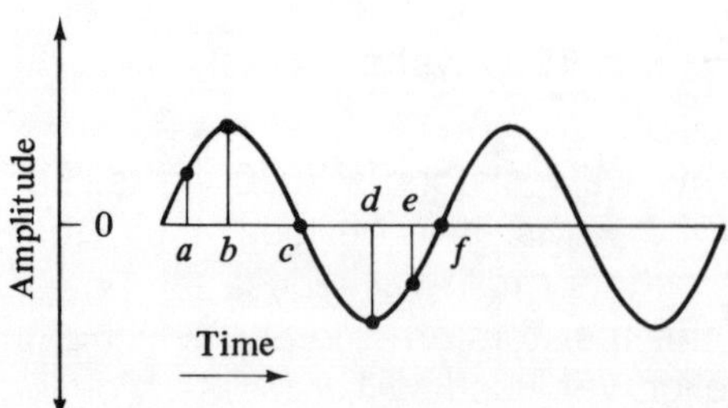

FIGURE 10.7   Different amplitudes of instantaneous current or voltage

RMS **10.11**

In discussing and using alternating currents and voltages it is necessary to have a convenient means of indicating amplitude. This presents a problem since an alternating voltage is constantly changing from zero to some maximum, returning to zero, reversing polarity to rise to an opposite maximum voltage, declining to zero, reversing polarity, and so on over and over again. One might use, for instance, the maximum amplitude attained. This is not as useful, however, as measuring peak-to-peak voltage (as is easily done with the oscilloscope explained in Section 8.11). One disadvantage of either of these methods is that they do not yield a value that can be used in the power formula to calculate the power dissipated (in heat) in a resistor. To achieve this end, an **effective voltage** usually called **rms (root-mean-square)** is used.

To understand effective voltage, examine Figure 10.8. When the voltage is zero, the power will be zero, ($p = e^2/R$). As the voltage rises to its peak, the power will increase in proportion to the square of the voltage. Note that even when the voltage is negative, the square is positive, and the power is always zero or higher. Naturally, the maximum power occurs when the voltage reaches its (positive or negative) peak (e_p). At that point the power is $e_p{}^2/R$. The power declines from this point to zero but rises again to a maximum when the voltage is in the opposite direction.

If a dashed line is drawn through the center of the wave representing power (at $e_p^2/2R$), it can be seen that if the amount above the line (shaded area) were cut off, it would exactly fill in the troughs below the line (also shaded). In other words, the dashed line was drawn at

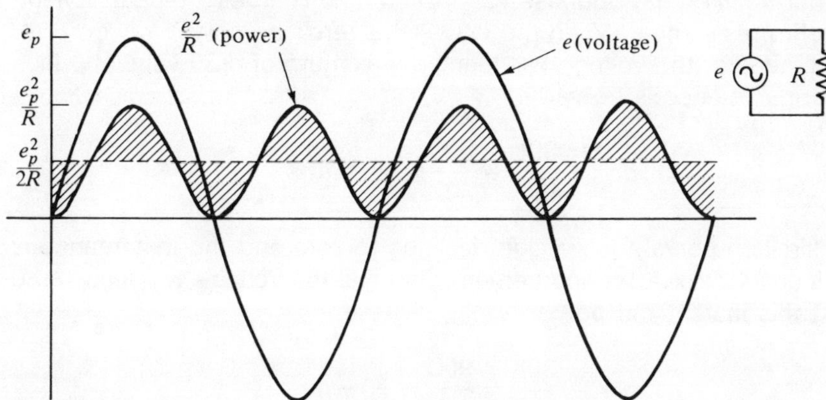

FIGURE 10.8   Power in a resistor due to an alternating voltage

the **average (mean)** power being dissipated. Since the maximum amplitude of the power wave was the peak voltage squared divided by the resistance ($p = e_p^2/R$), the mean value is half of this:

$$e_p^2 \times \frac{1}{2} = \frac{e_p^2}{2R}$$

If a direct-current voltage developed the same power in this same resistor, if would equal E^2/R. Therefore,

$$\frac{E^2}{R} = \frac{e_p^2}{2R}$$

If both sides of this equation are multiplied by R, the result is

$$E^2 = \frac{e_p^2}{2}$$

The direct-current voltage (**effective voltage**) that will cause the same amount of heat as e_p can be found by taking the square root of both sides of the equation:

$$E = \sqrt{\frac{e_p^2}{2}} = \frac{e_p}{1.414} = .707 e_p$$

Notice that when we established this value, we took the square root of the *mean* value of the square of the peak voltage. For this reason, the effective voltage, E, is called the *root-mean-square* or *rms voltage*. It is always .707 times the maximum voltage or half this amount (.3535) of the peak-to-peak voltage. For instance, if the power-line voltage that comes to your home is 117 volts (rms), the voltage peak is actually $117/.707 = 165$ volts, and the peak-to-peak voltage is 330 volts.

      Let us take an example of all of the foregoing using numbers rather than letters. Suppose we applied a 60-volt peak-to-peak voltage to a 45-ohm resistance. When the wave is at zero, the power will be zero. A short time later the voltage will reach a maximum of +30 volts at which time the instantaneous power is

$$p = \frac{(+30)^2}{45} = \frac{900}{45} = 20 \text{ watts}$$

After a similar interval the voltage declines to zero and the instantaneous power is again zero. After another short interval the voltage will reach −30 volts. At this instant the power is

$$p = \frac{(-30)^2}{45} = \frac{900}{45} = 20 \text{ watts}$$

It should now be obvious that the average power is midway between zero and 20 watts. Therefore, it is 10 watts. A direct current voltage that will develop 10 watts in a 45-ohm resistor is 21.21 volts.

$$p = \frac{E^2}{R} = \frac{(21.21)^2}{45} = 10 \text{ watts}$$

Note that we can arrive at the 21.21-volt figure by multiplying the peak voltage by .707.

$$30 \times .707 = 21.21 \text{ volts}$$

Root-mean-square is the most common way of specifying AC voltage or current. If no other indication is given, it should always be assumed that a current or voltage is in terms of rms.

From the previous discussion we have seen how resistance is important not only to control current but also to absorb power and dissipate it as heat. This unique characteristic gives us a method of expressing a varying voltage by a single number (the effective or rms voltage). Now, it is time to examine the common commercially available hardware specifically designed to have resistance.

FIXED RESISTORS 10.12

Section 3.7 introduced carbon resistors and the color coding that is commonly used. When higher precision is needed than can be provided by carbon-composition resistors, several other types are available. When the precision must be very high or if high power dissipation is needed, resistors are usually **wirewound.** High resistance wire is insulated (sometimes with just a coating of varnish) and wound on a tube or core of insulating material. Usually this assembly is encased in an airtight (**hermetically sealed**) plastic or ceramic coating.

In addition to wirewound resistors, fairly high precision can be obtained from **deposited-film** resistors. In this latter type, a glass (or ceramic) tube is coated with a fine spray of metallic or carbon particles. A helical groove resembling a screw thread is then ground from the surface so as to leave a spiral conducting path of the required resistance.

Although color codes are occasionally used on resistors other than carbon types, it is most common to find precision resistors with the resistance value, tolerance, and recommended maximum power dissipation printed on a protective jacket molded over the resistance element.

10.13 *VARIABLE RESISTORS*

In addition to fixed-value resistors such as have been described, there are also many commercial varieties of variable resistors. The most common of these are rheostats and potentiometers. In power applications rheostats are very common, but they are infrequently used in electronics.

A **rheostat** is usually a coil of resistance wire with a wiper blade or contact that can be moved over its length. Two attachment **lugs** (or contacts) are provided. One is electrically connected to the wiper and the other is connected to one end of the coil. When a short length of resistor wire is between the coil end and the wiper, the resistance is low. As the wiper is moved to make the amount of resistance wire between the lugs greater, the resistance is increased.

A **potentiometer** often closely resembles a rheostat in appearance. The difference is that three lugs are provided. The third lug is connected to the other end of the resistance wire (or surface).

Potentiometers for electronics work are frequently made by substituting for the resistance wire a surface coated with a resistive material such as carbon. The connections of a rheostat and of a potentiometer (the slang term is *pot*) are illustrated in Figure 10.9.

With the rheostat, a portion of its resistance is in *series* with the load. That is, all current flowing through the rheostat must also flow through the load resistor. In the case of the potentiometer, a portion of the resistance of the pot is in series with the load, but the remaining portion is in *parallel* with it. In other words, the current *divides* at the wiper so that part of it flows through the load and the other portion bypasses the load.

Of course, a potentiometer can be used as a rheostat merely by not connecting one of the end terminals, or by connecting it to the

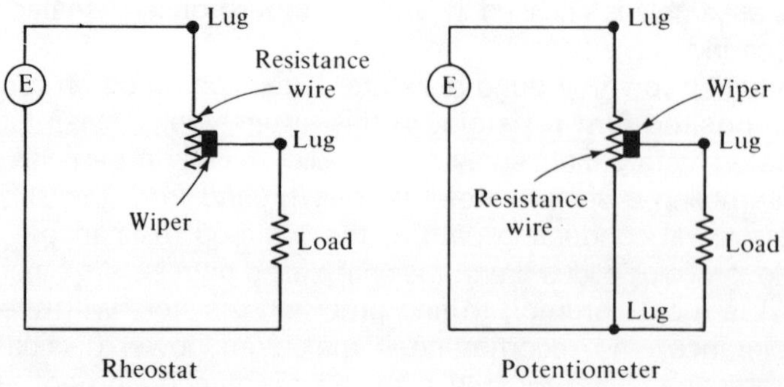

FIGURE 10.9 Variable resistor circuits

wiper terminal. Particular care must be exercised in adjusting potentiometers to a low resistance setting since the increase in current due to the
decreased resistance may cause too much power dissipation in the short
length of wire or coated surface. This will burn a spot on the surface, or
melt the wire, and render the potentiometer useless.

Variable resistors and potentiometers are illustrated in Figure 10.10.

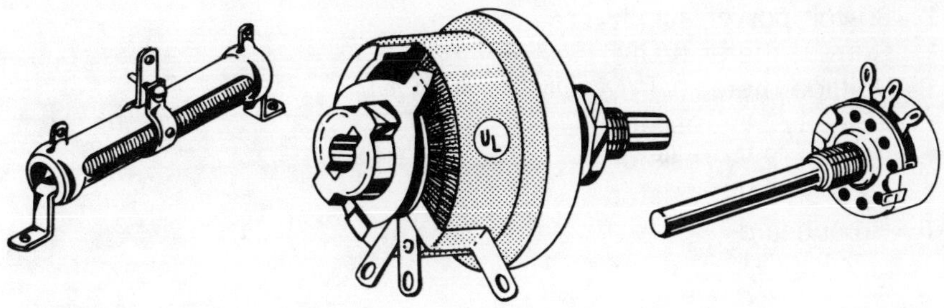

FIGURE 10.10 Variable resistors

CURRENTS AND VOLTAGES IN A SERIES-PARALLEL CIRCUIT

Equipment needed

- 1—30-volt power supply
- 1—Current meter (VOM)
- 1—Voltage meter (VTVM, DMM, or VOM)
- 1—15-K, 1/2-W resistor
- 1—22-K, 1/2-W resistor
- 1—33-K, 1/2-W resistor
- 1—Breadboard

PROCEDURE

1. Construct the circuit illustrated in Figure A.

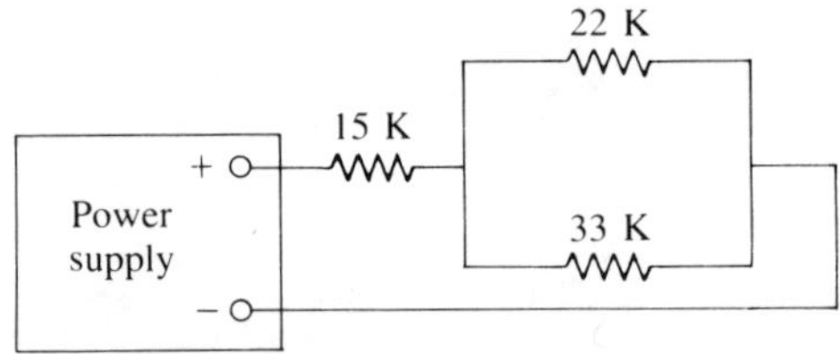

FIGURE A Series-parallel circuit

2. Insert the current meter in series with the 15-K resistor. Put the positive terminal of the meter on the positive terminal of the power supply and connect the negative terminal to the resistor.

3. Turn the power supply on and adjust it to 25 volts.

4. Observe and make note of the total current in the circuit.

5. Move the current meter so that it is in series with the 22-K resistor. Put the positive terminal of the meter at the junction of the 15-K and 33-K resistors, and connect the negative terminal to the 22-K resistor.

6. Observe and make note of the current through the 22-K resistor.

7. Move the current meter so that it is in series with the 33-K resistor.

8. Observe and make note of the current through the 33-K resistor.

9. Compare the total of the readings of steps 6 and 8 with the current measured in step 4. What law does this illustrate?

10. With a voltage meter measure the voltage across the parallel pair and subtract this value from 25 to establish the voltage across the 15-K resistor.

11. Divide the value calculated in step 10 by 15 K and compare to the value of current found in step 4.

12. Divide the voltage measured across the parallel pair by the current measured in the 15-K resistor and compare this with the resistance of the parallel pair as calculated by the product-over-sum formula.

OSCILLOSCOPE VOLTAGE MEASUREMENT

Equipment needed:

    1—Variac (variable transformer)
    1—Oscilloscope
    1—Voltage meter (VOM)
    1—6.8-K, 1/2-W resistor
    1—3.3-K, 1/2-W resistor
    1—Breadboard

PROCEDURE

1. Construct the circuit illustrated in Figure B.

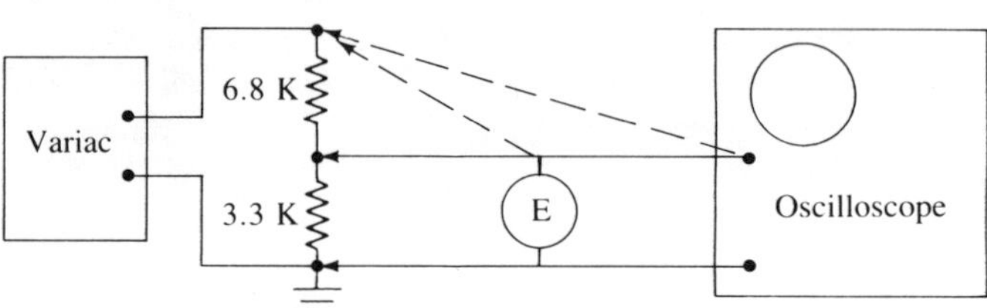

FIGURE B Oscilloscope voltage measurement

2. Adjust the variac so that 30 volts (rms) as measured by the VOM is applied across the total circuit.

3. With the VOM, measure the voltage across the smaller resistor.

4. With the oscilloscope, measure the peak-to-peak voltages across the smaller resistor and across the total circuit.

5. Compare the VOM readings with rms values calculated from the oscilloscope readings. (The peak-to-peak oscilloscope readings can be converted to rms by multiplying by 0.3535.)

POWER, RESISTANCE, AND RMS

QUESTIONS

From the following list select any item or combination of items (prefix, word, or number) to fill each of the blanks in the questions.

amperes	volts	22.2
capacitive	watts	25
carbon	wirewound	27
center	1/8	37
composition	0.20	45
deposited	0.23	47
distance	1/4	50
end	0.44	60
ER	1/2	65
E/R	0.51	72
E^2/R	0.89	77.8
film	1.0	84.6
IE	1.2	86.4
inductive	1.35	100
IR	2.0	147
I^2R	2.12	200
I/R	3.84	229
I^2/R	6.43	238
kilo	8.1	250
milli	8.8	403
micro	10.5	635
ohms	12	1050
rate	13.5	1200
reactance	15.6	19,050
resistance	21	

1. The total current in the circuit in Figure C is ___________.

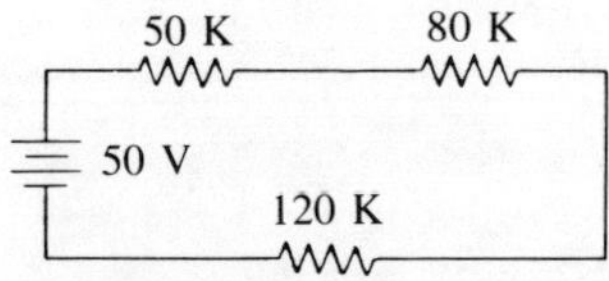

FIGURE C Series circuit

2. The total resistance of the circuit in Figure D is ___________.

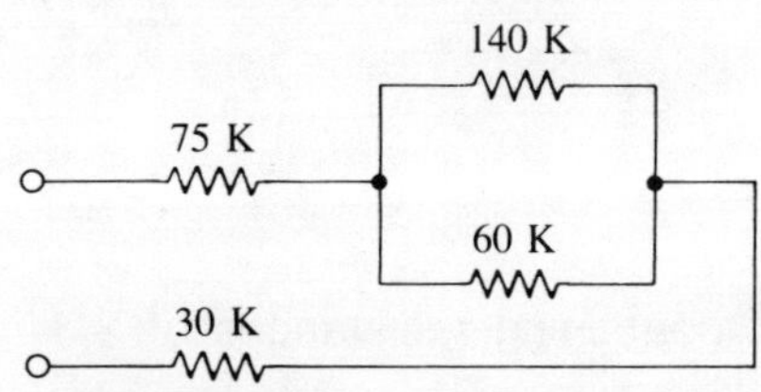

FIGURE D Series-parallel circuit

3. The voltage across the 8-K resistor in Figure E is _________.

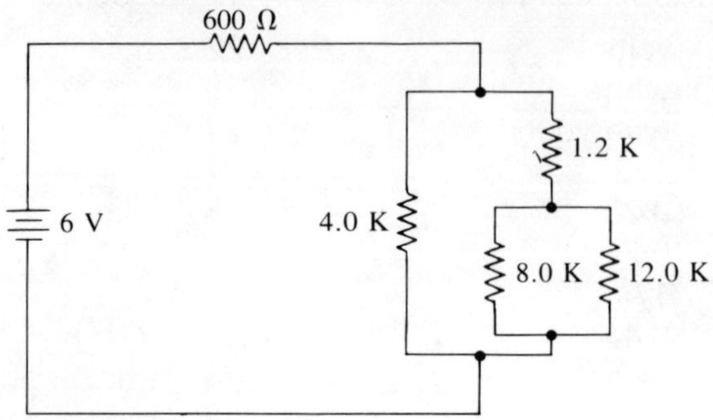

FIGURE E Complex series-parallel circuit

4. The current drain from the battery in the circuit shown in Figure F is _________.

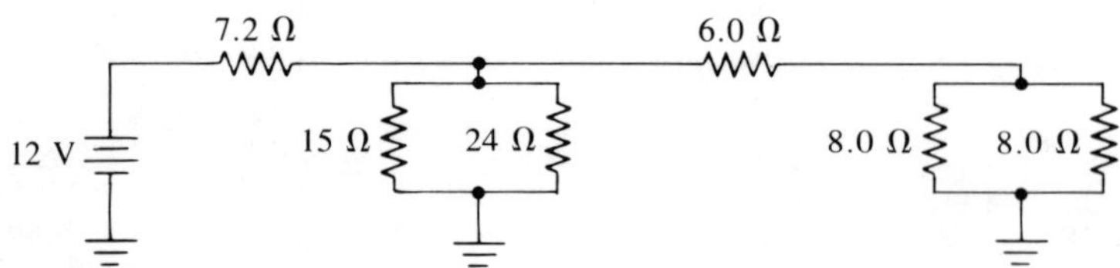

FIGURE F Circuit with common grounds

5. In the circuit in Figure G the current through the 15-K resistor is _________.

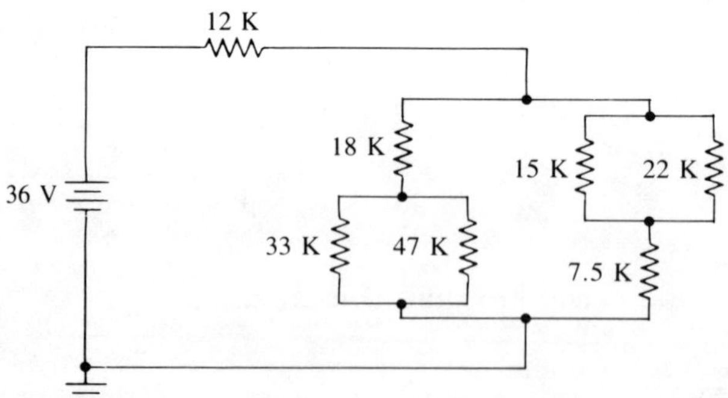

FIGURE G Complex circuit

6. The total resistance of a series circuit containing resistances of 3.3 kilohms, 15.0 kilohms, and 750 ohms is _________.

7. The resistance of a series circuit consisting of a 22-ohm resistor, a 33-ohm resistor, and a rheostat set at 10 ohms is _________.

8. A string of 9 Christmas tree lights has all the lamps in series. When plugged into a wall socket at 120 volts, the current is 600 milliamperes. The resistance of each bulb is __________.

9. Parallel resistors of 47 ohms, 63 ohms, and 27 ohms have a total resistance of __________.

10. An electric circuit contains a section consisting of parallel resistances of 470 ohms, 330 ohms, and 150 ohms. The equivalent resistance of the section is __________.

11. 120 volts is applied across a circuit in which a 36-ohm resistor is in series with a parallel combination of a 60-ohm resistor and a 40-ohm resistor. The total resistance of the circuit is __________.

12. In the circuit described in question 11, the voltage drop across the 36-ohm resistor is __________.

13. In the same circuit, (question 11) the current in the 40-ohm resistor is __________.

14. In the same circuit as the previous three questions, the applied voltage must be changed to __________ volts if the current in the 40-ohm resistor is to be changed to 1.0 ampere.

15. In the circuit shown in Figure H, when the resistors have the values shown, the total resistance is __________.

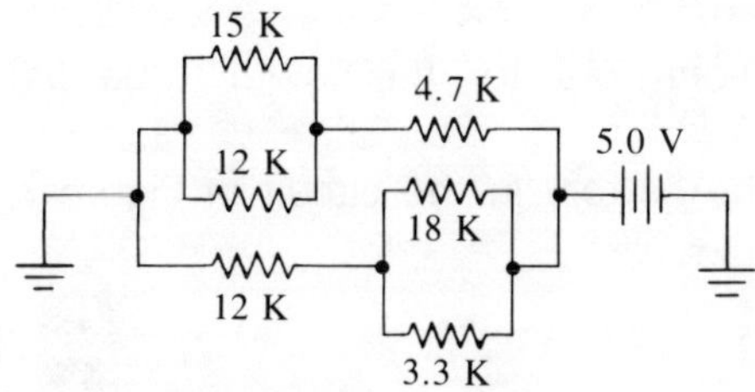

FIGURE H Complex circuit

16. In Figure I the current in the 12-ohm resistor is __________.

17. In the Figure I the current in the 40-ohm resistor is __________.

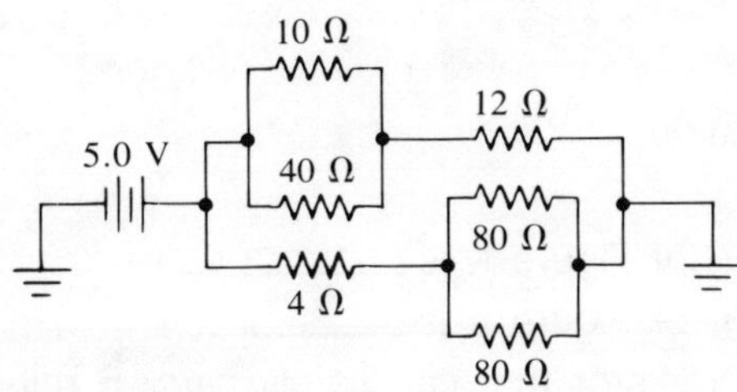

FIGURE I Complex resistor network

18. In Figure H, if the voltage applied is raised to 500 volts, the total current through the circuit is __________.

19. In the circuit shown in Figure J, the voltage at point *A* with respect to ground is __________.

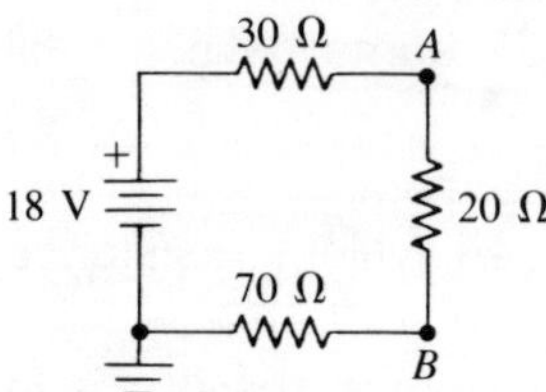

FIGURE J Series circuit

20. In the circuit in Figure J, the voltage at point *B* is __________.

21. Work can be defined as a force multiplied by the __________ through which it is applied.

22. Power is the __________ of doing work.

23. In electrical calculations, the power, measured in watts, is calculated by any of three formulas, as follows:

$$P = \underline{\quad\quad}$$
$$P = \underline{\quad\quad}$$
$$P = \underline{\quad\quad}$$

24. If 15 volts is placed across a 510-ohm resistor, the power to be dissipated by the resistor is __________.

25. The power being drained from the battery in the circuit in Figure K is __________.

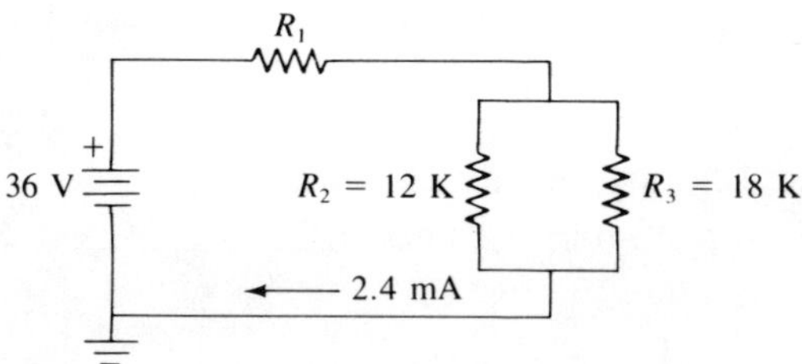

FIGURE K Power in a series-parallel circuit

26. The power being dissipated in R_2 of Figure K is __________.

27. The power being dissipated in R_1 of Figure K is __________.

28. The smallest standard-size 27-K resistor that can safely carry 6 milliamperes when well ventilated is a __________ watt size.

29. The resistors in Figure L are pictured actual size. (See Figure 3.3, pg. 37, for resistor size comparison.) Calculate the power that must be

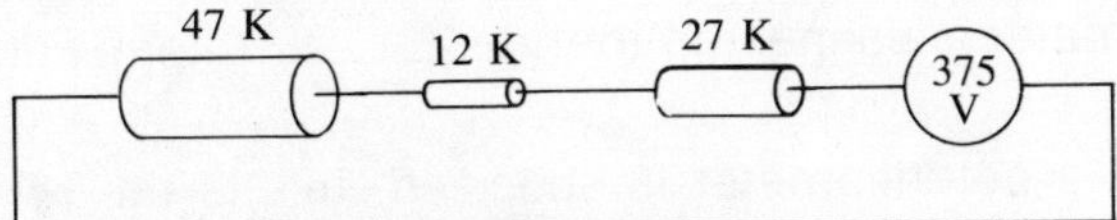

FIGURE L Critical resistor size

dissipated in each resistor. The ________ __________ resistor is the most likely to overheat.

30. Carbon resistors of 1/4 W, 1/2 W, 1 W, 2 W, 5 W, and 10 W are commonly stocked by suppliers. The smallest size that could be safely used if 35 volts is to be placed across 680 ohms is ________.

31. How much power does an electric iron take if the current is 8.75 amperes when the line voltage is 120 volts? $P =$ ________.

32. How much power must be dissipated by a 75-ohm resistor if 5.5 volts is placed across it? $P =$ ________.

33. A flashlight bulb has a resistance of 15 ohms when a current of 300 mA flows through it. How much power is used? $P =$ ________.

34. A technician has calibrated an oscilloscope so that a sine wave of 1.0 volt peak-to-peak makes a line 1.0 inch high on the CRT when the step attenuator is on *X1*. An unknown voltage causes the line to be 0.7 inch long when the attenuator is on *X100*. What is the rms voltage being measured? $V =$ ________.

35. An oscilloscope has been calibrated to show 1.0 V *p-p* as a line 2.0 cm high when on a *X1* vertical attenuator setting. When an unknown voltage is measured with a *X10* setting, the line is 4.6 cm high. The rms voltage of the unknown source must be ________ volts.

36. With a step attenuator set at *X1*, an oscilloscope is calibrated to display a 1.0-V *p-p* voltage as a vertical line 10 units in length. An unknown voltage subsequently input to the scope causes a display 6 units long when the step attenuator is at *X10*. The unknown rms voltage is ________ volts.

37. A 36-V *p-p* sine wave voltage is applied across a 680-ohm resistor. The power developed is ________ and the smallest standard-size resistor permissable is ________ watt.

38. With a VOM a technician measures the voltage across part of the circuit of a child's toy train powered by house current. His meter reads 6.2 volts. What is the peak voltage? $E_{pk} =$ ________.

39. A sine wave has a peak voltage of 324 volts. What is the rms value? $V =$ ________ volts.

40. A sine wave reaches a peak voltage of 22 volts. What rms voltage is this? $E_{rms} =$ ________ volts.

41. When very high precision is required, resistors are usually ________. However, fairly high precision is obtained by ________ ________ resistors.

42. The wiper of a potentiometer is connected to the __________ terminal (lug).

43. When the resistance of a potentiometer is specified, this is the resistance between the __________ lugs.

44. Power in the form of heat is developed only in the variety of impedance called __________.

11

Meters

GENERAL 11.1

While it is not intended that this book be more than a quick survey of the field of electronics, the prime tool of a technician—the meter—must not be ignored.

METER CONSTRUCTION 11.2

The common meter is really a tiny electric motor that operates against a spring and stalls when the spring tension is reached that it cannot overcome. The angle through which the meter has turned is then a measure of the current flowing through its motor winding. Unlike an ordinary motor, the meter is not designed to spin around. In fact, it must not be rotated beyond about 90° where pins stop the motion of the needle attached to the **rotor** or **armature**.

Good laboratory meters use the d'Arsonval movement, which has jeweled bearings like a fine watch.

The heart of the meter consists of a permanent magnet bent in a sort of horseshoe shape and a pivoted coil (the armature) that can rotate in the magnetic field between the two poles. (See Figure 11.1.)

A needle is attached to one end of the armature coil, with spiral springs on both ends. The springs serve two purposes. They are attached to the ends of the winding of the coil and provide for passing current through it. Also, they are attached to the frame of the instrument so that when there is no tension on the springs, the needle is at the zero position of a scale placed under it. The armature assembly is mounted in the gap between the soft-iron poles of the permanent magnet. When direct current is passed through the coil, a magnetic field is created that is opposed by the field from the permanent magnet. The opposition of the fields forces the armature to turn. This means that the needle moves until the twisting of the spiral springs is just enough so that the tension of the springs is equal to the field opposition. Of course, the more current that flows in the coil, the more the needle turns. The scale under the needle is **calibrated** to read in standard units.

To use a meter to make measurements of alternating current, it is necessary to place diodes in the circuit so that current will flow through the meter in one direction only. The inertia of the armature will not allow the needle to follow a changing current. Instead, it will move to the position of average current. This point is not at .707 of the peak current on the direct-current scale, and meter manufacturers are forced to make scale corrections in order to provide rms readings.

Meters are made for many different applications and ranges of values. As a result, some have many turns of very fine wire in the armature while others have fewer turns or larger wire. For measuring current, an ideal **ammeter** would carry any amount of current and have no

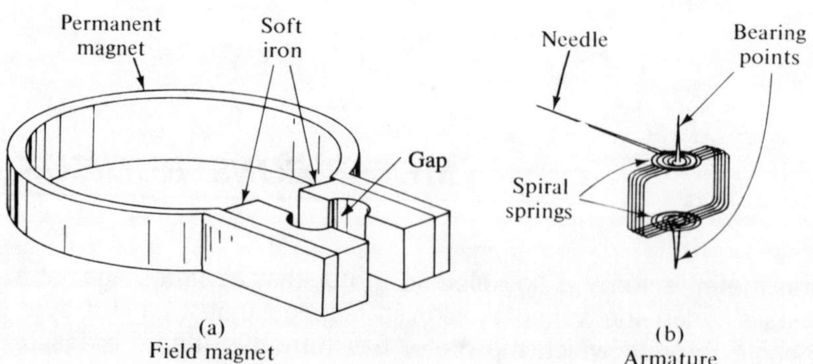

FIGURE 11.1 Principal meter parts

resistance. In this way it would not disturb the circuit into which it was inserted. This imaginary meter is impossible since the armature winding must obtain power from the circuit being measured. Consequently, current limitations and resistances vary from meter to meter. It is quite common to find meter movements that are designed to have a full-scale reading when a 1-milliampere current flows through them, but one model might have 25 ohms resistance and another 250 ohms. Meter movements are readily available that will give full-scale readings from 20 microamperes to 30 milliamperes.

EXTENDING SCALES 11.3

A meter movement can be used to measure currents that are much higher than its full-scale value by the simple expedient of bypassing the meter coil with a **shunt** resistor designed to allow only a fraction of the total current through the coil. This method is illustrated in Figure 11.2.

In the circuit illustrated, the shunt resistor R_s is placed in parallel with the meter. The total circuit current I_T will divide so that one part of it (I_m) will flow through the meter coil and the other part (I_s) will flow through the shunt. If we consider I_m to be the current that will cause the meter to read full scale, then the voltage drop across the shunt resistor must be ($I_s R_s$). This must be equal to $I_m R_m$ because this is a simple parallel circuit and the voltage drop from A to B is the same whether we take a path through the meter or through the shunt. Putting this in an equation and dividing each side by I_s we find that

$$R_s = \left(\frac{I_m}{I_s}\right) R_m$$

However, the shunt current I_s is the total current I_T less the meter current.

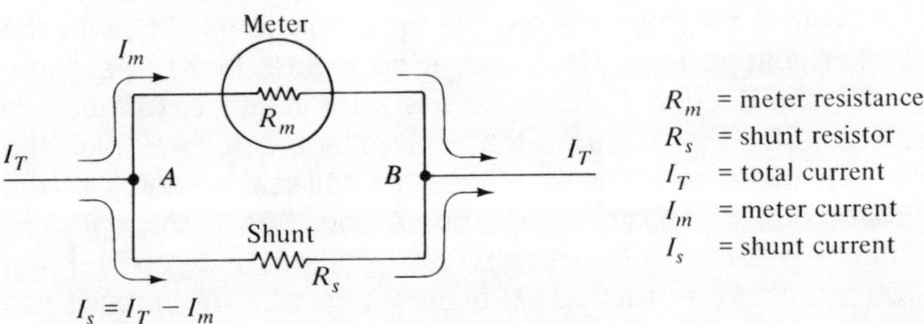

FIGURE 11.2 Meter range extended

$$I_s = I_T - I_m$$

Replacing I_s in our former equation with this value, we have

$$R_s = \left(\frac{I_m}{I_T - I_m}\right) R_m$$

As an illustration of the use of this formula, consider the following problem: A meter having an internal resistance of 48 ohms and a full-scale deflection when carrying a 1-milliampere current is to be used to measure currents as high as 5 milliamperes. What must the shunt resistance be? Substituting these values into the equation, we have

$$R_s = \left(\frac{.001 \text{ A}}{.005 \text{ A} - .001 \text{ A}}\right) 48 \ \Omega$$

$$= \left(\frac{.001 \text{ A}}{.004 \text{ A}}\right) 48 \ \Omega$$

$$= 12 \ \Omega$$

Naturally, the scale under the needle should be modified to provide convenience in reading the meter. This modification would consist of a renumbering so that five major divisions would cover the space from the zero on the left to the full-scale point on the right.

When using a current meter care should be taken that the meter resistance is negligible compared to the other resistance in the circuit. It should be obvious, for instance, that a meter having an internal resistance of 50 ohms would cause the current in a circuit of 150 ohms to be greatly reduced so that the meter reading would be erroneous.

11.4 *MEASURING VOLTAGE*

A meter movement can also be used to measure voltage by placing a resistor in series with it and then placing the assembly in parallel with the item across which voltage is to be found. (See Figure 11.3.) The same meter movement as in Section 11.3 can be used, for instance, to measure voltage as high as 100 volts. Assume that the voltage across R_s plus the voltage drop of the meter is 100 volts. Since the full-scale reading of the meter is 1 milliampere, I_m equals 1 milliampere and .001 A $(R_s + R_m) =$ 100 V. Divide both sides of this equation by .001 ampere and you find that $R_s + R_m = 100 \text{ V}/.001 \text{ A} = 100000 \ \Omega$. Since R_m is only 48 ohms it can be considered of no consequence (compared to R_s). Therefore, a series resistor of 100 kilohms will convert the 1-milliampere meter to a voltmeter.

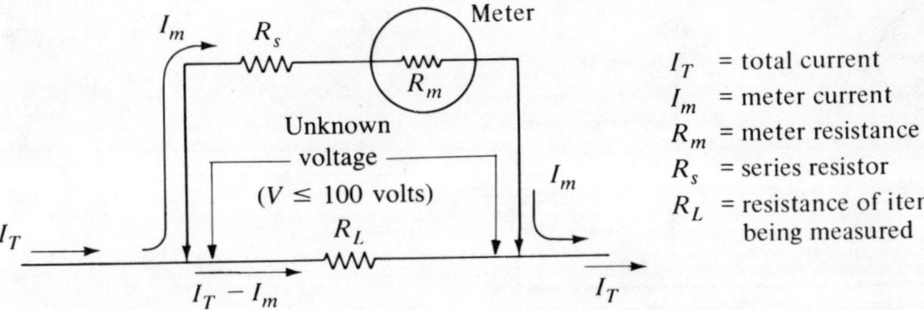

FIGURE 11.3 Meter movement as a voltmeter

Again, of course, the scale should be modified to reflect the proper values between zero and full-scale. An ideal voltmeter would be just the opposite of an ideal ammeter. That is, it would have infinite resistance so that it would not disturb the circuit being measured. As with the current meter, however, this ideal is practically impossible since some current must be robbed from the circuit to drive the armature of the meter. Very sensitive meters can be made that require very little current and, therefore, can have very high resistance. Most meters have a label that indicates the voltmeter resistance by the term **ohms per volt**. For instance, if a meter marked 1000 ohms per volt has its selector switch set on the 10-volt scale, the internal meter resistance is 10 × 1000 = 10,000 ohms. Switch such a meter to the 100 volt-scale and the internal resistance will be 100 × 1000 = 100,000 ohms. A meter is not satisfactory for measuring voltage across a resistance of more than one-tenth the meter resistance because the meter is in parallel and would seriously lower the resistance of that portion of the circuit.

MEASURING RESISTANCE 11.5

Several methods use a meter movement to measure resistance, but the most common is the series **ohmmeter**. This consists of a meter movement, a battery, a fixed-value series resistor and a variable series resistor as shown in Figure 11.4. The battery in the circuit furnishes the current necessary to power the meter. There must be no other voltage source in the circuit to be measured when using the ohmmeter. When the probes X and Y are an open circuit the meter needle will be at the zero-current position. This is interpreted as the infinite resistance reading of the ohmmeter. When the probes are closed together, the battery will cause the meter needle to move across the scale. The variable resistance R_p

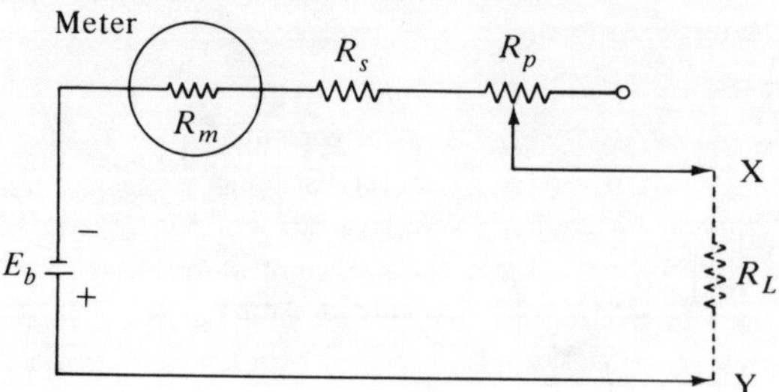

FIGURE 11.4   Series ohmmeter

should then be adjusted so that the needle will read full-scale. This point must be considered to be zero resistance (the resistance between the probes). Any other value of resistance (R_L) placed between the probes after this adjustment will cause the needle to take a position on the scale between the two end values. The scale must, therefore, be recalibrated to read these values. The scale will not be **linear**. That is, the size of the graduation for 1 ohm on the left end of the scale will be very small but on the right end will be fairly large.

11.6 *LIMITATIONS*

Meters such as the ones previously described have the advantages that they are quite portable and require no warm-up before use. In addition, an ammeter or **VOM** (combination ammeter, voltmeter, and ohmmeter) will measure current directly by placement of the meter in series with the circuit.

In spite of these excellent qualities, some disadvantages exist. For a meter to operate, it must take some power from the circuit being measured. Naturally, this disturbs the circuit somewhat. As stated in Section 11.4, this is a serious disadvantage with a voltmeter if the meter resistance (including the series resistor) is not large compared to the resistance across which the voltage is measured. For instance, if a meter with a resistance of 100 kilohms were placed across a resistance of 400 kilohms, as in Figure 11.5, the voltage reading would be erroneous.

In Figure 11.5 the meter resistance is in parallel with the 400-kilohm resistor. The resistance of the combination can be calculated by the product-over-sum formula as follows:

$$\frac{(100 \times 10^3)(400 \times 10^3)}{(100 \times 10^3) + (400 \times 10^3)} = \frac{4 \times 10^{10}}{5 \times 10^5} = 80 \text{ K}$$

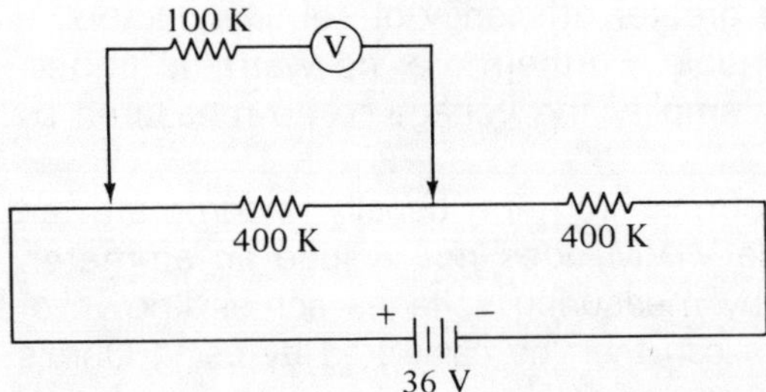

FIGURE 11.5   Relatively small voltmeter resistance

The meter would show the voltage to be appropriate for this resistance.

$$I = \frac{E}{R_T} = \frac{36}{(400 \times 10^3) + (80 \times 10^3)} = \frac{36}{480 \times 10^3} = 75 \ \mu A$$

$$V = IR = (75 \times 10^{-6})(80 \times 10^3) = 6 \ V$$

It is obvious from the circuit illustrated that the voltage across the resistor should be half the supply voltage (18 volts) since it would divide between the two equal resistors.

From the foregoing it follows that an ideal voltmeter would have an infinite resistance. Such an ideal device is practically impossible, but fortunately very-high resistance voltmeters can be made that derive their power from an external source.

THE VTVM AND ELECTRONIC METERS 11.7

As discussed in Section 5.12, a charge (voltage) placed on the control grid of a vacuum tube can control a large current flowing from cathode to anode (through the tube). This control-grid bias can be derived from a very small current flowing through a very large resistance (typically 10 megohms or more), and the resulting current through the vacuum tube can be placed in series with a meter. This principle has been used to create the **vacuum-tube voltmeter (VTVM).** The power to drive the meter movement comes from an external source (usually rectified house current). Such meters can be built with a power pack (batteries) to make them portable, but this adds weight and cost. In addition, several minutes of warm-up time are required for proper operation. However, this meter has much less loading effect on the voltage to be measured than a VOM.

Since the advent of semiconductors, **electronic meters** have been developed that employ the same basic principles as the VTVM but utilize transistors. Because of the greater efficiency of semiconductors, it is easier to make these units portable. Furthermore, no warm-up is necessary. All these meters basically amplify the voltage to be measured by a fixed amount.

VTVM's and electronic meters usually incorporate an ohmmeter in their design, but the VTVM does not include an ammeter. Current readings must be made by measuring voltages across known or measured resistances. The currents can then be calculated by using Ohm's law.

11.8 *THE DIGITAL MULTIMETER (DMM)*

During the seventies, the sensational development of integrated circuits, light-emitting diode and liquid-crystal displays, and refined printed circuits made digital meters easily portable and relatively inexpensive. (See Figure 11.6.) Digital meters replace the needle-and-scale readout with a numerical

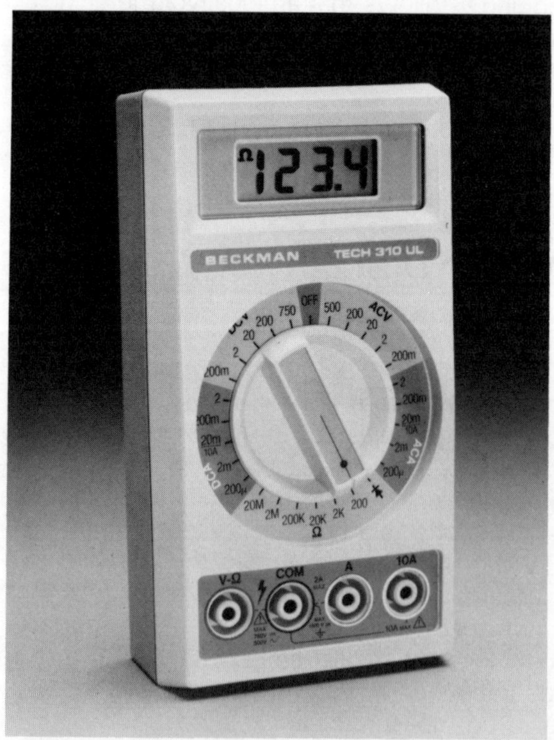

FIGURE 11.6   Digital multimeter (courtesy of Beckman Instruments Inc.)

readout (display). Not only are they easy to read, but because they have high input impedance when set to measure voltage, they are excellent for making measurements across high resistances. In versatility, they surpass the VOM since they can measure current, voltage, or resistance, but with greater ease and precision.

THE WHEATSTONE BRIDGE 11.9

Another method of measuring resistance is the use of a **Wheatstone bridge.*** This circuit provides much more accurate resistance measurements than an ohmmeter, but it is more cumbersome, costly, and time consuming. A typical Wheatstone bridge circuit is illustrated in Figure 11.7.

A meter movement is used that has its zero point in the middle of the scale. When current flows in one direction the needle will move to the right, but if it is reversed the needle will move to the left. R_1 and R_2 are fixed-value resistors of known but different values. R_v is a variable resistor whose resistance is known for each setting (a dial is provided on a potentiometer). R_x is an unknown resistor to be measured by the bridge. When the circuit is connected as shown, the resistance of R_v is adjusted until the meter reads zero. At this point the voltage A must be the same as at B or current would flow through the meter. This means that the voltage drop across R_x must be equal to the voltage across R_1. Also, the voltage across R_v must equal the voltage across R_2. Since voltage drop is the product of current and resistance, $I_a R_x = I_b R_1$ and $I_a R_v = I_b R_2$.

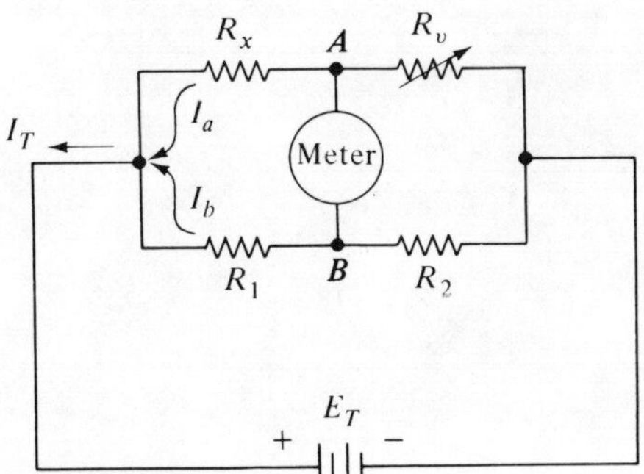

FIGURE 11.7 Wheatstone bridge

*Named for Sir Charles Wheatstone (1802–75) who popularized its use.

Therefore,

$$\frac{R_x}{R_1} = \frac{I_b}{I_a} = \frac{R_v}{R_2}$$

Since $R_x/R_1 = R_v/R_2$,

$$R_x = \left(\frac{R_1}{R_2}\right) R_v$$

R_i and R_2 are known values, and R_v can be read from a dial. A simple calculation will, therefore, tell us the value of R_x. In commercial instruments of this type, the ratio of R_1 to R_2 is incorporated in the calibrations of the R_v dial so that it reads directly in ohms for the resistance of R_x.

METER RESISTANCE MEASUREMENT AND SHUNT CALCULATION

Equipment needed

- 1—Power supply (adjustable)
- 1—10-K potentiometer
- 1—100-ohm potentiometer
- 1—1-mA meter movement
- 1—Current meter (VOM or DMM)
- 1—Wheatstone bridge (or other accurate resistance-measuring instrument)

PROCEDURE

1. Build the circuit illustrated in Figure A.

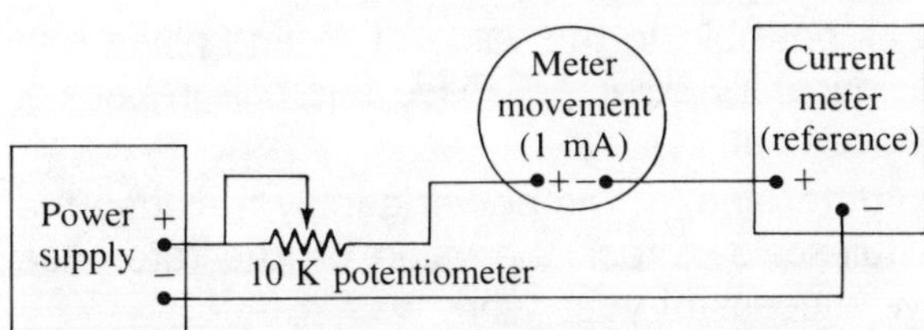

FIGURE A Measurement of meter full-scale current

2. Adjust the 10,000-ohm potentiometer to its full resistance, then turn on the power supply and adjust the voltage to allow enough current to flow so that the 1-mA meter shows a small deflection.

3. Gradually reduce the resistance of the 10-K potentiometer until the 1-mA meter reads full-scale.

4. Make careful note of the reading of current on the reference current meter. (This will be a reference reading for use in the next step.)

5. Place the 100-ohm potentiometer in parallel with the 1-mA meter as shown in Figure B.

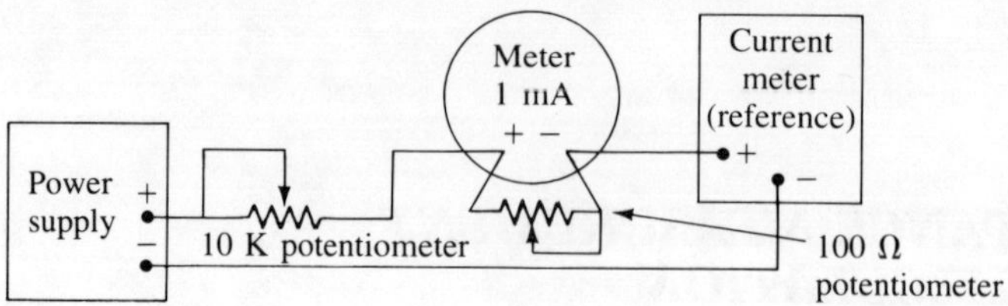

FIGURE B Matching of meter resistance

6. Simultaneously adjust both potentiometers until the 1-mA meter reads 1/2 scale while the current meter remains at the reference reading.

7. Carefully remove the 100-ohm potentiometer from the circuit without changing its setting and measure its resistance on a Wheatstone bridge or other accurate instrument. This resistance equals that of the 1-mA meter movement.

8. Calculate the shunt needed to allow the 1-mA movement to measure 10 mA. (See Section 11.3.)

9. Select a resistor (R_1) slightly larger than this value (R_2), measure its value accurately, then calculate the amount of resistance (R_x) to put in parallel with it by use of the following formula:

$$R_x = \frac{R_1(R_2)}{R_1 - R_2} \quad \text{where} \quad \begin{aligned} R_1 &= \text{selected resistor} \\ R_2 &= \text{desired shunt resistance} \\ R_x &= \text{approximate value of desired} \\ &\quad \text{parallel resistor} \end{aligned}$$

10. Find a resistor as close as possible to the value of R_x and place it in parallel with the resistor originally selected. Place this pair in parallel with the 1-mA meter movement.

11. Adjust the 10-K potentiometer and/or the power supply to cause the shunted meter to read full-scale. Check the value on the reference meter to determine how successful your work was.

BUILD A WHEATSTONE BRIDGE

Equipment needed

1—Decade resistance box
1—VTVM or DMM
1—1000-ohm, 1/2-W precision resistor*
1—10-K, 1/2-W precision resistor*
1—10- to 10,000-ohm precision resistor*
1—Power supply, or two dry cells

PROCEDURE

1. Construct the circuit illustrated in Figure C.

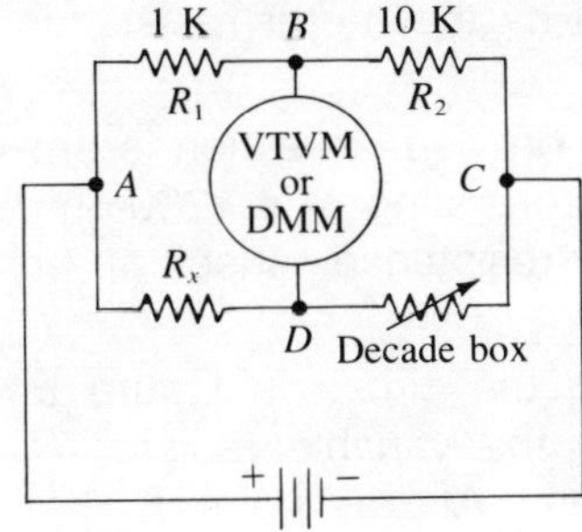

FIGURE C Rudimentary circuit

2. Set the VTVM on a low-voltage scale, short the leads together, and adjust the zero knob until the needle points to the center of the scale. (If a DMM is used this step is unnecessary.)

3. Place the VTVM or DMM in the circuit as shown in Figure D.

FIGURE D Wheatstone bridge

*The values of the precision resistors are not critical. The listed values merely make calculations easy. It is important, however, to **know** the values within 1 or 2 percent.

4. Connect the power supply or drycells from point *A* to point *C* in the circuit, and adjust the decade resistance until the VTVM needle (or DMM readout) indicates zero voltage.

5. Note that the decade resistance has been set to a value as many times larger than R_x as the resistor R_2 is to R_1. In this case, R_1 is 1000 ohms and R_2 is 10,000 ohms, so R_x is one-tenth as large as the decade resistance.

QUESTIONS

Select a symbol, word(s), number(s) or combination from the following list to fill each of the blanks in the questions.

current	negligible	0.00	500
DMM	ohms	1.58	1422
high	voltage	2.25	2844
kilohms	VOM	11.5	3000
low	VTVM	78	5688
millivolts		225	

1. Loss of magnetism in the field magnet of a d'Arsonval meter would cause the meter to read too __________.

2. The voltage drop across a 156-ohm meter movement when it is indicating full-scale at 500 microamperes is __________ __________.

3. To allow the meter movement in the previous question to be used to measure currents as high as 50 mA, a parallel shunt of __________ ohms must be used.

4. The meter movement of Question 2 is used to measure the current in a circuit consisting of a 12-volt battery in series with a total circuit resistance of 30 kilohms. The error caused by the meter resistance is __________.

5. The same meter movement as used in the previous question is employed to measure the current in a circuit consisting of a 1200-ohm resistance powered by 4.0 volts. The meter resistance caused an error of __________ percent.

6. A Wheatstone bridge is created by wiring the circuit in Figure E. An unknown resistor is used for R_x, and the variable resistor R_v is adjusted to cause meter *M* to read zero. At this point R_v has a resistance of 22.5 ohms. The resistance R_x is then known to be __________ ohms.

   METERS

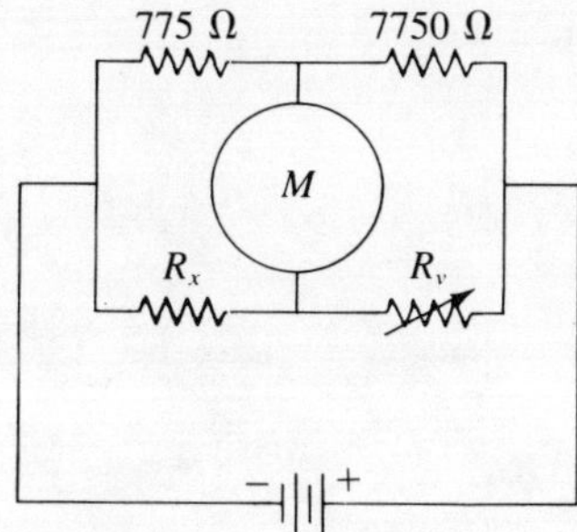

FIGURE E Typical bridge circuit

7. A VOM has the advantage over a VTVM that it can directly measure __________.

8. High internal impedance is important for meters used to measure __________. Two general types of meters that have high impedence are the __________ and the __________.

9. To permit a meter movement such as that specified in question 2 to be used to measure a voltage as high as 250 volts, a series resistor of __________ __________ must be used.

10. A meter movement such as specified in Question 2 is to be used with a 1.5-volt cell to create a series ohmmeter. A series resistance of 2844 ohms causes a full-scale reading with the probes connected together. The resistance between the probes that will cause a half-scale reading is __________ ohms.

12

Inductors and Transformers

INDUCTANCE **12.1**

About 150 years ago at the University of Copenhagen, Professor Hans Christian Oersted was demonstrating the properties of the electric battery, which had been invented about twenty years earlier. Lying on the table, for intended use in another demonstration, was a magnetic compass. When the professor connected a wire across the terminals of the battery, the wire accidentally lay across the compass, and to his amazement the compass needle moved. He cautiously made no mention of his observation to his class but experimented with a larger battery and conductor at a later time. What he had found was that any conductor is surrounded with a magnetic field whenever current flows in it. The field is strongest at the surface of the conductor but gets weaker and weaker farther away from it. Also, the greater the current flow, the stronger the field becomes. If a circuit is constructed with a battery, variable resistor, and conductor, the current can be made very small by making the resistance large. Or, the current can be increased by reducing the resistance. Such a circuit is illustrated in Figure 12.1.

165

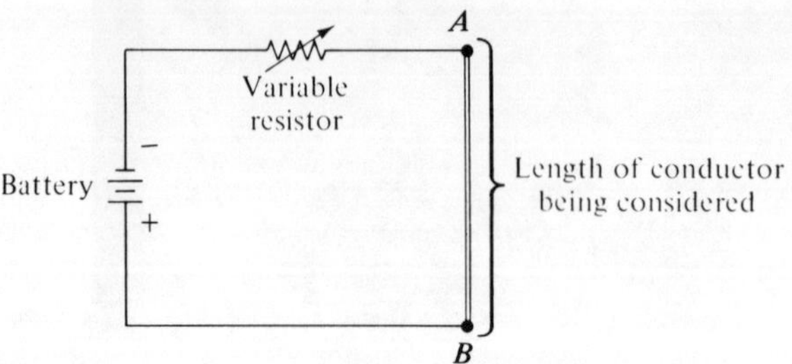

FIGURE 12.1   Circuit with varying current

Consider that portion of the conductor between points *A* and *B*. For purposes of illustration the magnetic field around the conductor is shown in Figure 12.2 by a varying density of dots. A high density indicates a strong field and a low density a weak one.

This illustration shows that when the current is strong, the density of the field is as great some distance away from the conductor as it is right next to the conductor when the current is weak. If the current is gradually increased from weak to strong, it appears that the field strength next to the conductor has moved outward. For this reason, it is generally stated that an increase in current causes the magnetic field to move outward. It naturally follows that reduction of current causes the field to move inward, or collapse.

In addition to the fact that a change in current causes the magnetic field to move, it is also true that if a magnetic field is moved across a conductor in a closed circuit, a current will flow in the conductor. For instance, consider the circuits shown in Figure 12.3.

Conductor segments *dc* and *ef* are very close together. When the switch is closed so that current starts to flow from *d* to *c*, the expanding magnetic field will move across conductor *ef*. While it is moving, there will be a surge of current from *f* to *e*. However, as soon as the

FIGURE 12.2   Effect of current on magnetic field

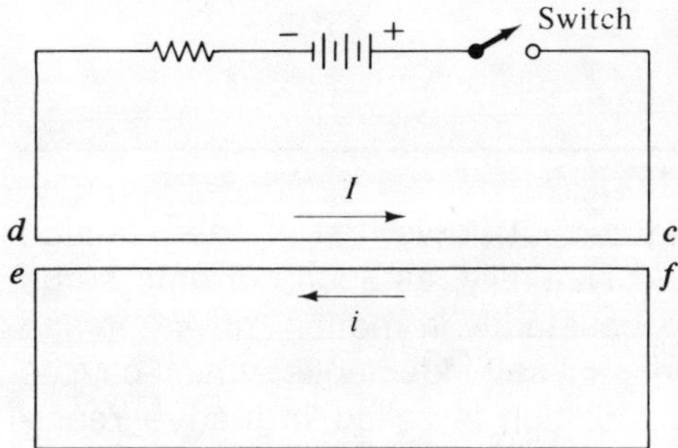

FIGURE 12.3 Inductance of current in a closed secondary circuit

current from *d* to *c* reaches its maximum value, the magnetic field will no longer move and the current from *f* to *e* will stop. If the switch is opened again, the current from *d* to *c* will cease. This will cause the magnetic field to collapse back into the wire, *dc*. The movement back across *ef* will, again, cause a current surge, but this time it will move from *e* to *f*.

The action just described that causes current to flow in a second closed circuit (adjacent to, but not electrically connected to the first), is dependent on **coupling**. Coupling means that the magnetic field of a conductor can move across another conductor and induce EMF.

SELF-INDUCTANCE 12.2

A single conductor behaves as though it is a pair of conductors that have been moved so close together that they are merged. As an expanding or contracting field moves through the outer portions of a wire, a voltage is created to oppose the change in current causing the magnetic field movement. This effect is called **self-inductance** and is measured in a unit called the **henry*** (abbreviated *H*).

The *henry* is defined as the inductance value that will cause a counter EMF of one volt if the current changes at the rate of one ampere per second.

---

*Named for Joseph Henry (1799–1878), a U.S. physicist who did much of the developmental research concerning electromagnets and relays.

12.3 *INDUCTIVE REACTANCE*

As discussed previously, an alternating current flows first in one direction and then reverses to flow in the other. Naturally, as such currents surge to and fro in a wire, a magnetic field is constantly in motion moving in and out. This field movement results in continual self-inductance, which causes an opposition to the AC flow. This opposition is called **inductive reactance**, which is a form of impedance and is therefore measured in ohms. The faster the current is forced to change, the greater the reactance becomes. Recall that in Section 6.7 an analogy was suggested that relates inductance to the inertia of mass. Consider, therefore, how much more difficult it is to shake a heavy object rapidly than it is to push it slowly to and fro. This is analogous to changing the current direction frequently in an inductor rather than slowly and gently. It follows, then, that the higher the frequency of the AC, the more the reactance will be. If the waveform of the applied voltage is a pure sine wave, the reactance can be calculated by the following formula:

$$X_L = 2\pi fL$$

where

X_L = inductive reactance in ohms

π = 3.14

f = frequency in Hertz

L = inductance in henries

Ohms's law applies to an inductor by substituting inductive reactance for resistance in the familiar formulas.

$$I = \frac{E}{X_L}$$

$$E = IX_L$$

$$X_L = \frac{E}{I}$$

12.4 *INDUCTORS*

Self-inductance occurs in any wire to some extent as described in Section 12.2. However, the inductance can be increased by the simple tactic of

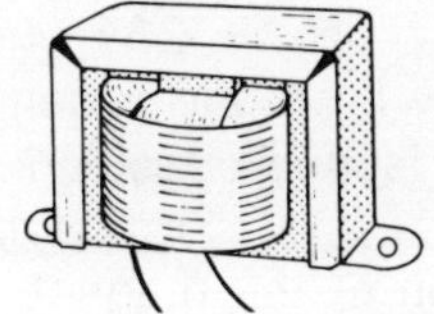

FIGURE 12.4 Fixed-value inductors

winding the wire into a coil. When this is done, the magnetic field moving to or from one turn of the coil will cross adjacent turns. This action of each turn upon all the others causes a large voltage in opposition to a current *change*. The effect can be even further increased by providing an easy path for the magnetic field. Iron is an excellent material for this purpose and the majority of inductors, therefore, have iron cores. Where frequencies are great, ferrite cores (special powdered iron) are used to prevent losses caused by currents that may be induced in the core itself. When very-high frequencies rule out even ferrite cores, the iron is omitted.

Commercial inductors are made in many sizes and shapes, with various types of construction. Typical fixed-value inductors are shown in Figure 12.4.

Variable inductors are sometimes used that consist of a coil with a movable iron core. As the core is moved into the coil, the inductance is increased. When it is moved out, the inductance is decreased.

It should be observed that a common term applied to one application of an inductor is **choke**. The origin of this term can be understood if one visualizes the effect that an inductor would have on a current that is a mixture of AC and DC. The higher the frequency of the AC, the greater the inductive reactance to oppose that portion of the current. In effect, the AC would tend to be "choked off" and the DC would be unaffected.

INDUCTANCES IN SERIES 12.5

When inductors are placed in series, their inductances add together. Since the letter L is used to represent inductance, the formula for this is

$$L_T = L_1 + L_2 + L_3 + \ldots + L_n$$

The formula just stated assumes that no coupling exists. When two inductors are coupled, additional inductance is created called **mutual inductance**. This is measured in henries just as self-inductance and is represented by the letter M. The result of two coupled inductances is then

$L_T = L_1 + L_2 + 2M$ when the interaction is additive (series aiding), or $L_T = L_1 + L_2 - 2M$ when the interaction is in opposition to the self-inductances (series opposing).

From what has just been shown, one can see that the direction of the magnetic fields is very important since it may increase or decrease the total inductance.

Since inductive reactance is proportionate to inductance, the reactances in series merely add together to give total reactance. Where there is no coupling, the formula for this is

$$X_{L_T} = X_{L_1} + X_{L_2} + X_{L_3} + \ldots + X_{L_n}$$

12.6 INDUCTANCES IN PARALLEL

As you might expect from the fact that inductances in series followed the same pattern as resistors, the formulas are similar for parallel connection. A reciprocal relationship applies.

$$\frac{1}{L_T} = \frac{1}{L_1} + \frac{1}{L_2} + \frac{1}{L_3} + \ldots + \frac{1}{L_n}$$

and

$$\frac{1}{X_{L_T}} = \frac{1}{X_{L_1}} + \frac{1}{X_{L_2}} + \frac{1}{X_{L_3}} + \ldots + \frac{1}{X_{L_n}}$$

12.7 TRANSFORMERS

When two conductors lie parallel (side-by-side), a change in the current in one induces a voltage in the other, which will cause current flow if the second conductor is part of a closed circuit. This fact is the basis for one of the most useful of all electrical components—the **transformer**. This device was briefly described in Section 6.8. (See Figure 12.5.)

The induction of a current in one conductor because of the change of current in the other is due to the movement of the magnetic field. This effect can be **enhanced** (made greater) by winding the coils of each conductor very closely together so that all the magnetic field from one conductor must pass through the other. (This technique is called **tight,**

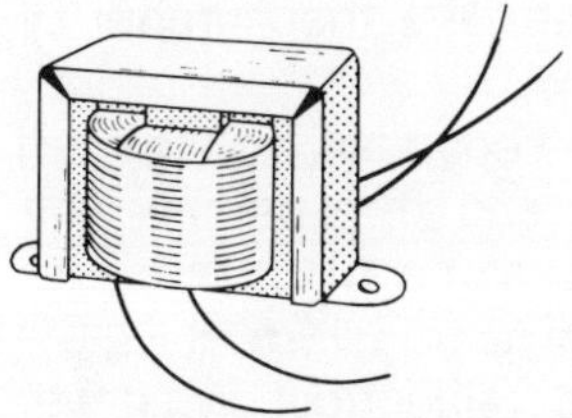

FIGURE 12.5 Typical transformer

unity, or **close coupling**.) Enhancement is further achieved by providing a path for the magnetic field of low **reluctance** (opposition to presence of a magnetic field). A vacuum or air provides extremely high reluctance, but iron has low reluctance (or resistance to magnetic **flux**). Transformers are commonly wound on an iron core that is in a shape that provides a closed magnetic loop.

It is an important fact that the number of turns of wire used for each conductor has a significant effect upon the relationships of voltages across each coil and the currents that will be induced in them.

First, a fairly large number of turns in each coil is needed to assure good coupling (that is, to permit the greatest interception of the magnetic field by the turns in the other coil winding).

If the coupling is complete (100 percent), the voltages across the windings will be in proportion to the *ratio* of the number of turns. *Ratio* means the result (quotient) from dividing one number by the other. In other words, if 15 volts AC is placed across a **primary winding** (so called because it is the one used for input current) of 100 turns, a voltage of 45 volts will be found across a **secondary winding** (output winding) of 300 turns. If the secondary winding has 400 turns, the voltage across it will be 60 volts.

Anyone who has used a hydraulic jack knows that a relatively small force applied repeatedly at the handle causes the jack to apply the large force necessary to lift an automobile. The principle by which the jack operates is analogous to the operation of a step-up transformer. The jack handle is linked to a small diameter piston that pumps fluid into a large diameter cylinder. It is easy to push the small piston because the force from the handle is distributed over the tiny area of the small piston. With many strokes, the high-pressure fluid builds up under the large piston causing the very large upward force. The large piston does not move far with each jack handle stroke, but the force is far greater than that applied to the small piston because the same pressure is applied over a much larger area. In this analogy, the force on the jack handle is like input voltage. The total distance through which the handle moves is like input current. The output force lifting the automobile is equivalent to the transformer's output voltage, and the distance the car is lifted is analogous to the secondary current. The jack piston diameters are like the number of turns on the primary and secondary windings.

In summary, with a transformer, the secondary voltage will be proportionate to the secondary-to-primary turns ratio. The primary current will be the secondary current multiplied by the same turns ratio.

One of the most important uses of a transformer is the ability to obtain an alternating voltage of a desired value from whatever voltage is available merely by selecting the appropriate turns ratio.

Another use for the transformer is as an **isolation** device. The primary and secondary windings have no electrical connection, but the signal from one induces the signal in the other.

Still another use for the transformer is a rather subtle one. It is a fact of electronics that the maximum transfer of power from one component to another occurs when their impedances are equal (matched). A transformer can be used to make unequal impedances appear to be matched.

A transformer does not transform current. However, the current in the primary will be governed by the load in the secondary. For instance, consider the (previous) transformer with a 100-turn primary and a 300-turn secondary, and assume that a load of 5000 ohms is placed across the secondary. The 45 volts in the secondary winding will cause a secondary current of 9 mA (I = 45V/5000 = .009 A). This means that the power dissipated in the load is 405 mW. [P = EI = (45 V) (.009A) = 405 mW.] If there are no losses, the input *power* will be identical and will be the product of the input voltage and current. Since the primary voltage is one-third that of the secondary, the current required for the same power is three times as much or 27 milliamperes. [P = (15 V) (.027A) = .405 W.]

Suppose that a device having a resistance of 40 ohms is to be used to power a circuit having a resistance of 1000 ohms. A transformer having a primary of 100 turns and a secondary of 500 turns is used between the two devices. If the primary voltage is 20 volts, the secondary will be 100 volts and the current that will flow in the secondary will be 100 ÷ 1000 = 0.1 A. The primary current will, therefore, be five times as much or 0.5 A. Since the primary voltage is 20 volts, the load appears to the power supply as being 20 ÷ 0.5 = 40 ohms. This amounts to a sort of electronic "trickery" to make the impedances match. It is done by making the turns ratio equal to the square root of the impedance ratio to be matched.

Transformers are generally very efficient devices, but, as with any component, there are some losses. It is axiomatic that P_{in} = P_{out} + P_{losses} (power-in is equal to power-out plus losses).

12.8 *AUTOTRANSFORMER*

A valuable variation of the transformer called the **autotransformer** does not isolate the primary and secondary windings from one another. A single

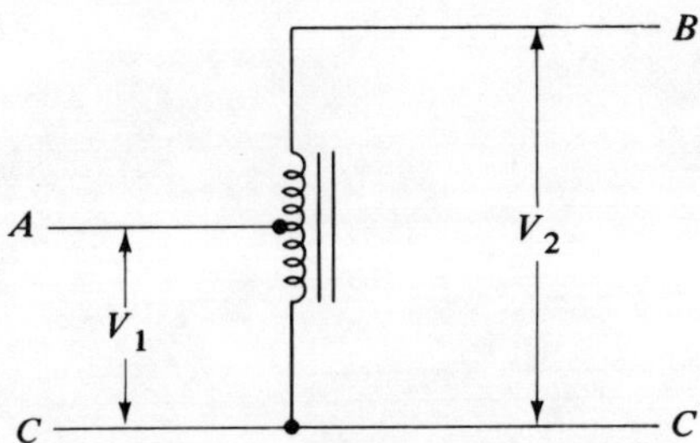

FIGURE 12.6 Autotransformer

coil is wound on a core and a tap is provided to a point along the coil as shown in Figure 12.6. In effect it is an AC inductive voltage divider.

If the leads A and C are connected to an input alternating voltage V_1, current flowing through the length of coil between them will create a changing magnetic field in the iron core around which the entire winding is wound. The output voltage V_2 between B and C will be proportionate to the ratio of the turns between B and C divided by the turns between A and C.

By connecting lead B to a sliding contact on the length of the coil, the voltage between B and C can be made adjustable.

INDUCTIVE TIME CONSTANT 12.9

In Section 6.7 it was pointed out that when a DC voltage is applied to an inductor there is a delay in the flow of current caused by the counter EMF generated by the expanding magnetic field. The degree of delay is affected by the resistance of the circuit. To illustrate this, consider the circuit in Figure 12.7 consisting of a battery, switch, inductor, and resistor. The max-

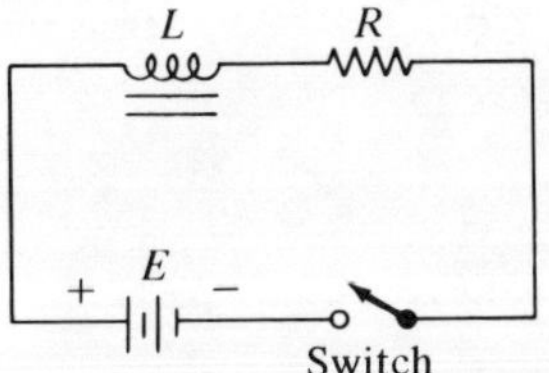

FIGURE 12.7 Circuit with inductor and DC voltage

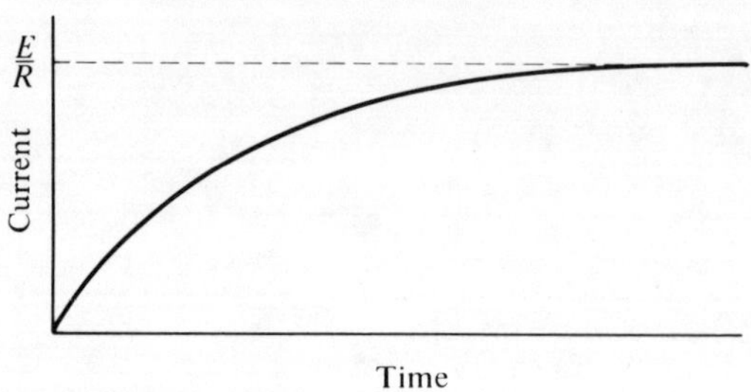

FIGURE 12.8   Effect of counter EMF on current flow

imum current that can flow in the circuit is that determined by Ohm's law: $I = E/R$.

When the switch is closed the delay caused by the counter EMF from the inductor allows the current to increase gradually to the E/R value as shown in Figure 12.8.

Suppose the inductance L is 10 henries, the resistance R is 100 ohms, and the voltage E is 12 volts. Then, the graph of the current will appear as shown in Figure 12.9.

If the resistance is increased to 200 ohms, the current approaches maximum much faster, as indicated in Figure 12.10.

The rate at which the current rises in such a circuit is also dependent on the inductance. Increasing inductance has an effect opposite to that of increasing resistance. The rate of change of current decreases as inductance is increased. In other words, the current takes longer to build up when the inductance is large. The effect of both inductance and resistance on the time needed for current to reach maximum can be stated as follows: *Time is proportionate to inductance and inversely proportionate to resistance.*

This fact gives us a means of expressing the inductive characteristics of a circuit without the necessity of drawing a curve as in

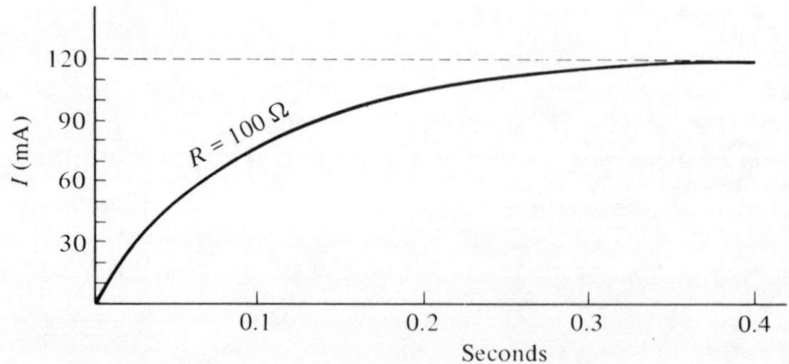

FIGURE 12.9   Effect of resistance on delay of current

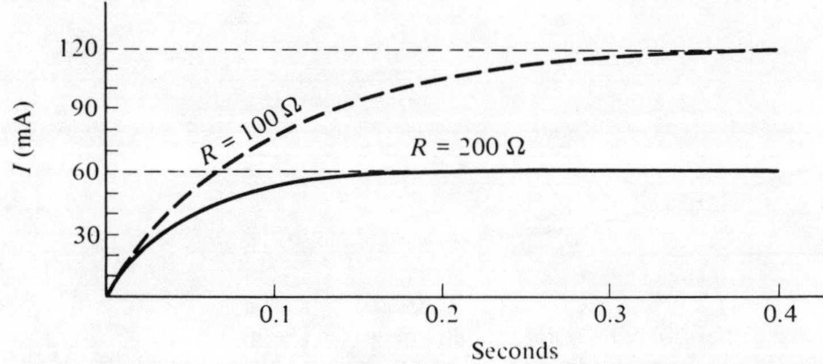

FIGURE 12.10 Current change with counter EMF delay

Figures 12.8, 12.9, and 12.10. If the inductance (in henries) is divided by the resistance (in ohms) the result is the number of seconds required for the current to change by an amount that is 63.2 percent of the ultimate amount it *can* change. For instance in the circuit shown in Figure 12.7 with a 12-volt battery, an inductance of 10 henries, and a resistance of 100 ohms, this calculation would yield 1/10 second. What this means is that from the moment the switch is closed to 1/10 second later, the current will rise from zero to 63.2 percent of the ultimate E/R value or,

$$.632 \left(\frac{12}{100} \right) = .07584 \text{ A} \cong 76 \text{ mA}$$

The calculated ratio (in the above case, 1/10 second) is called the **time constant**. If we represent the time constant with the symbol τ_L* the formula is stated simply as

$$\tau_L = \frac{L}{R}$$

Remember that τ_L is the time in seconds—from *any* instant—necessary for current to change 63.2 percent of the amount it can ultimately change. Consider the change in current in the previous circuit starting 1/10 second *after* closing the switch. First, understand that the ultimate change is the E/R value less the amount by which the current changed in the first 1/10 second. The ultimate change is, therefore, 12/100 − 0.07584 = 0.04416 ampere. In this second 1/10-second interval after the closing of the switch, the current will change by 63.2 percent of 0.04416, or 0.02791 ampere. Thus, the current 2/20 second after closing the switch will be the current at 1/10 second plus the change in the next 1/10 second for

---

*τ is a Greek letter pronounced "tau" to rhyme with "law."

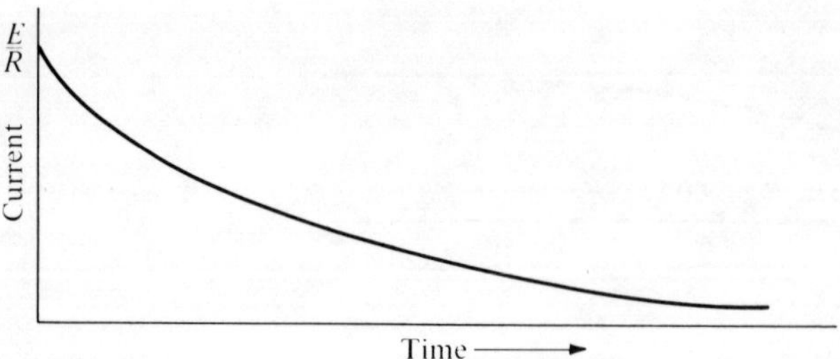

FIGURE 12.11   Current change with sudden halt of applied voltage

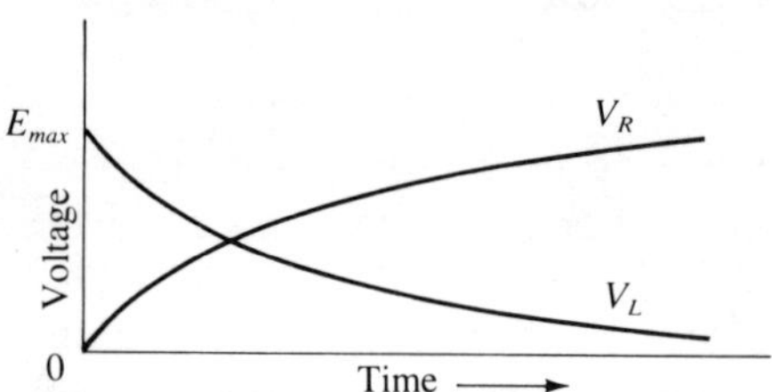

FIGURE 12.12   Effect of time on component voltages

a total of $0.07584 + 0.02791 = .10375$ ampere $\cong 104$ milliamperes. Note that the current after 2/10 second is approximately 104 mA on the graph in Figure 12.9.

Theoretically, the current will continually increase forever as it approaches the E/R value. As a practical matter, it can be assumed to be at the E/R value after about five time constants.

If, after the current has been flowing for some time, the source voltage is removed, the collapsing field around the inductor will cause current to continue to flow at a declining rate. For instance, suppose the battery of Figure 12.7 is suddenly shorted after the switch had been closed for several minutes. The graph of current will then be as shown in Figure 12.11.

The rate of reduction can be calculated as before, in terms of the time constant. It will drop 63.2 percent in the first period of L/R seconds, 63.2 percent of the remainder in the next period of L/R seconds, and so on.

Figure 12.9 has illustrated the increase in current with the passage of time after a DC voltage is placed across a circuit of an inductor in series with a resistor. Figure 12.12 shows how the voltage V_L across the inductor will fall with time while the voltage V_R across the resistor will rise.

PRACTICAL TIME CONSTANT 12.10

As seen by the preceding, the ratio of L to R provides the time constant of an inductor in series with a resistance. Unfortunately, this time constant yields the time for current to change by the unhandy amount of 63.2 percent. For convenience it would be nice to know the time needed for the current to change by the simple round number of 50 percent. $L/R \times .7$ yields such a time value and is referred to as the **practical time constant.***

LEFT-HAND RULES 12.11

No discussion of inductance would be complete without consideration of the **left-hand rule**, which relates conductor movement, magnetic field, and electron flow. This rule is the basis for the operation of generators, alternators, motors, electromagnets, and various other useful devices.

If the left hand is held as shown in Figure 12.13, the middle finger represents a conductor that is being moved across a magnetic field in the direction indicated by the thumb. If the pointer finger points in the direction of the magnetic field (from N to S), electron movement will occur in the direction pointed by the middle finger.

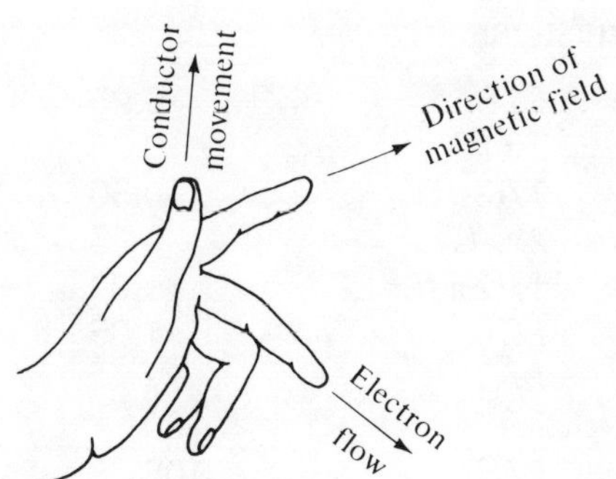

FIGURE 12.13 Left-hand rule for electron flow

---

*The figure .7 is an approximation. To 4 decimal places the value is .6931.

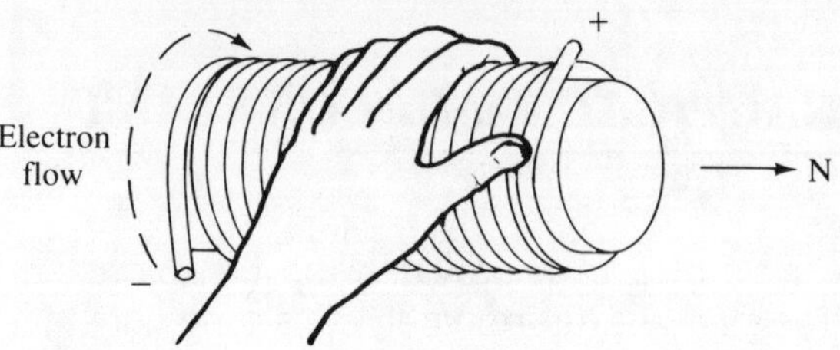

FIGURE 12.14 Left-hand rule for a coil

Another left-hand rule that is convenient relates the direction of current flow to the direction of the magnetic field around the conductor. The instruction for this is very simple. Place the thumb on the conductor pointing in the direction of the electron flow. Then, wrap the fingers around the conductor. The fingers point in the direction of the magnetic field.

Still another left-hand rule can be used with coil of wire. If the hand is wrapped around the coil so that the fingers point in the direction of electron flow, the thumb will point in the direction of the magnetic north pole. (See Figure 12.14.)

QUESTIONS

Select a letter, symbol, word, or number from the following list to fill each of the blanks in the questions.

ampere	milliwatt	C/R	0.25
current	of	$2\pi fC$	0.35
difference	ohms	$1/(2\pi fC)$	0.50
farads	over	$2\pi fL$	0.95
henries	product	$1/(2\pi fL)$	1.0
kilohms	reciprocal(s)	LC	1.05
no	sum	L/C	22
microampere	the	LR	100
microvolt	volts	L/R	108
microwatt	voltage	RC	417
milliampere	watts	R/C	450
millivolt	C/L	X/R	948

1. When a counter EMF of 1 volt is created by a rate of increase of 1 ampere per second the value of the inductor is __________ henry.

2. If a rate of increase of 4 amperes per second is required to create a counter EMF of 1 volt, the inductance is ___________ henry.

3. Inductive reactance is measured in ___________.

4. Inductive reactance of a sine wave is calculated by the formula: ___________.

5. An inductance of 12 henries is placed in series with a resistance of 24 ohms. One time constant after 36 volts is placed across the circuit, a current of ___________ mA has been reached.

6. The time constant for the circuit in the question 5 is ___________ seconds.

7. The time for the current to reach half of its ultimate value in the circuit of question 5 is ___________ seconds.

8. A transformer has a primary winding of 56 turns. The secondary contains 16 turns. When 77 volts is applied to the primary, the voltage across the secondary is ___________ volts.

9. When uncoupled inductances are in series, their total equivalent inductance is calculated as the ___________ of the inductances.

10. A 12-henry inductor is placed in parallel with an 8-henry inductor across a 3316-Hertz, 45-volt sine wave generator. The total reactance is ___________ and the current is ___________.

11. An inductor opposes a change in ___________.

12. Inductance presents ___________ opposition to steady-state direct current.

13

Capacitors

Since Chapters 10 and 12 reviewed resistance and inductance, only one final form of impedance, **capacitive reactance,** remains for our study.

As discussed in Chapter 6, capacitance exists when two conductive surfaces are separated by a **dielectric.** The dielectric can be any material (including air) that qualifies as an **insulator.*** If two separated conductive surfaces (or **plates**) have a DC voltage applied across them, electrons quickly flow from the negative terminal onto one surface and are drained (or repelled) from the other plate into the positive terminal of the DC supply. After this occurrence, no further current will flow because the plates are insulated from each other and also are **charged** with a voltage that is equal and opposite to the supply voltage.

---

*Insulators are those materials that require very large electrical forces to dislodge electrons from their parent atoms. Even when electrons are freed by such a force, there is little opportunity for them move into a nearby atom to dislodge another electron. In other words, insulators are materials that have extremely high (practically infinite) resistance.

181

There are three factors that determine the **capacitance** of a capacitor. Remember from Chapter 6 that capacitance is measured as the number of coulombs* that can be stored per volt applied. Increasing the surface area of the opposing plates will increase the capacitance, as will decreasing the spacing between the surfaces as long as the plates are still insulated from each other. The third factor affecting capacitance is the selection of the dielectric material used to separate the plates.

In Chapter 6 the elementary characteristics of a capacitor were discussed and capacitance was defined. Some repetition is necessary at this time to permit greater depth of study.

All capacitors store charge. It is tempting to relate this phenomenon to holding air in a pressure tank. Some similarity exists, such as the fact that the amount of air in a tank is dependent on the pressure applied to it, just as the amount of charge in a capacitor depends on the applied voltage. The maximum voltage that can be applied to a capacitor without permanent damage resembles reasonably well the maximum pressure an air tank can tolerate. In a capacitor this is a limitation of the **dielectric strength.**

However, nothing about the pressure tank seems to relate to two phenomena concerning capacitors. The capacitance of a capacitor can be increased by decreasing the thickness of the dielectric separating the conductive surfaces. Capacitance can also be increased by proper selection of dielectric materials. Any solid insulator is an improvement over a vacuum, and most are better than air. The multiple by which a dielectric material is able to increase the capacitance of plates with a given surface area as compared to using a vacuum for the insulator is called the **dielectric constant.**

Perhaps a closer analogy to a capacitor can be found in a balloon or hot-water bottle with a skin that stretches. If the skin material stretches easily, the amount of air that can be stored in the balloon at a given pressure is much greater than would be possible if the skin material were less flexible. The selection of a highly elastic material for a balloon skin would be similar to the selection of a material with a high dielectric constant for a capacitor. Further, a thin skin will stretch more easily than a thick one. There is quite a difference between inflating a child's balloon with a lung full of air and trying the same with a hot-water bottle. Both might be made of the same rubber material, but the thickness of the skin of the bottle makes inflation difficult. In this regard the capacitor is similar, because the thinner the dielectric (the closer the plates) can be made, the easier it is to store a large charge at a low voltage.

The larger the surface areas of the plates are (increasing the capacitance) the greater the instantaneous current will be at the moment voltage is applied, since more electrons are required to create the

---

*One coulomb equals 6.25 $\times$ 10^{18} electric charges (or electrons).

same degree of crowding on the negatively charged surface. If an alternating voltage is applied across the two plates, they will be alternately charged (and discharged) positive and negative. Again, the larger the surfaces, the greater the current will be before the surfaces reach the maximum supply voltage. Also, the more frequently the current is reversed, the greater will be the percentage of time in each cycle that electrons will be surging back and forth. As a result, the average current will be greater as the frequency of the applied AC is increased. Since current will be flowing in the conductors leading to the plates whenever the supply voltage is rising or falling, it can be said that alternating current "flows through" a capacitor. It must be realized, however, that electrons do not pass through. When an electron goes on to one surface, another leaves the opposite surface. Thus, the *charge* has moved through the capacitor but not the electron.*

CAPACITIVE REACTANCE **13.2**

From the preceding discussion it should be apparent that the flow of alternating current through a capacitor will be greater when the capacitance is greater or when the frequency is increased. Conversely, it is also true that the lower the frequency and/or the capacitance, the greater the impedance to alternating current. This form of impedance is called *capacitive reactance* and can be calculated for a sine wave source by the following formula:

$$X_c = \frac{1}{2\pi\, fC}$$

where

X_c = capacitive reactance in ohms

π = 3.14

f = frequency in Hertz

C = capacitance in farads

Note that this formula has a reciprocal relationship to the one for inductive reactance. Ohm's law continues to apply for pure capacitive reactance.

**Current is technically defined as rate of change of charge.*

$$I = \frac{E}{X_c}$$

$$E = I\,X_c$$

$$X_c = \frac{E}{I}$$

13.3 *CAPACITOR TYPES*

Fixed capacitors produced commercially are extremely varied. Two major types exist: those that can be used without regard for polarity, and **electrolytic capacitors** where one terminal must be negative or neutral at all times with respect to the other (positive) terminal. If correct polarity is not maintained, an electrolytic type will be severely damaged.

Capacitors that are not sensitive to polarity are made in many shapes and sizes, but their values of capacitance are usually quite small. One common type of construction consists of two strips of metal foil layered with a kraft paper separator (dielectric) that is rolled into a cylindrical package as in Figure 13.1.

Many different varieties of dielectrics are now used, such as mylar, mica, and ceramic. Cost is one consideration in the choice of a dielectric, of course. Ability to retain constant characteristics with changes in temperature is another. Commonly, the most important features are dielectric constant and dielectric strength.

Modern electronic designs almost always put a premium on space, so the ability to put a large capacitance in a small package is very desirable. The capacitance can be increased by increasing the area of the opposed surfaces, but this tends to enlarge the unit. The capacitance can be increased also by decreasing the distance between the surfaces (making the dielectric thinner), but this also decreases the amount of voltage the capacitor can withstand. However, some dielectrics can stand much higher voltages than others per unit of thickness. This characteristic is their dielectric strength. Thus, selection of a dielectric requires the evaluation

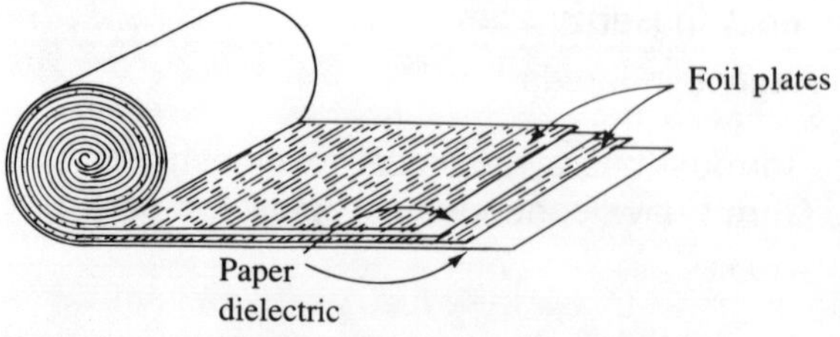

FIGURE 13.1 Paper capacitor construction

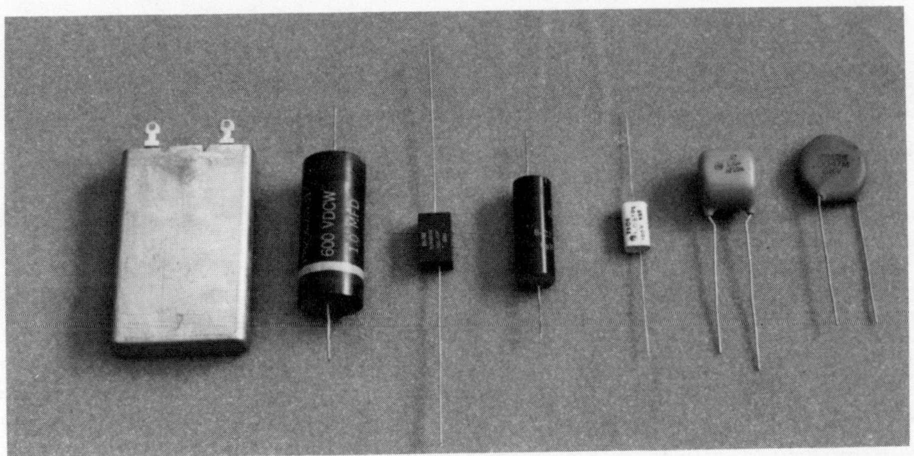

FIGURE 13.2 Fixed capacitors

of many factors. Different kinds of nonpolarized capacitors are illustrated in Figure 13.2.

Where very large capacitance is needed in a small package, **electrolytic** capacitors are frequently used. This capacitor consists of a strip of foil placed in a conductive paste (the **electrolyte**) that reacts with the surface of the foil when a small leakage current flows. The oxide formed creates a very thin insulator that performs as a dielectric between the conductive paste and the foil. The result is a very compact, high-capacitance device, although it is inherently "leaky" (in that it does not block DC current perfectly). Since it is also a diode, if the polarity is inadvertently reversed, high current will flow causing (at the very least) damage to the capacitor. In some cases, the effect is more serious, and the capacitor may explode like a small bomb. Several types of electrolytic capacitors are illustrated in Figure 13.3.

Variable capacitors are commonly made with air as the dielectric. Such components are usually employed in the tuning circuits of radios or transmitters. A **stator** consisting of a series of parallel, semicircular plates is insulated from the shaft of a **rotor** consisting of another set of alternately meshing semicircular plates. As the rotor is turned, its set of parallel plates slides in between the stator plates so that the area where the sets coincide can be increased or decreased. This action increases and decreases the capacitance of the device. (See Figure 13.4.)

Another type of variable capacitor often used is called a **trimmer**. This consists of two small surfaces with a dielectric between them (usually mica) and an insulated screw that can be turned to squeeze the plates together. The capacitance can be varied by very small amounts in this manner. The trimmer is generally used for making adjustments that will be changed infrequently.

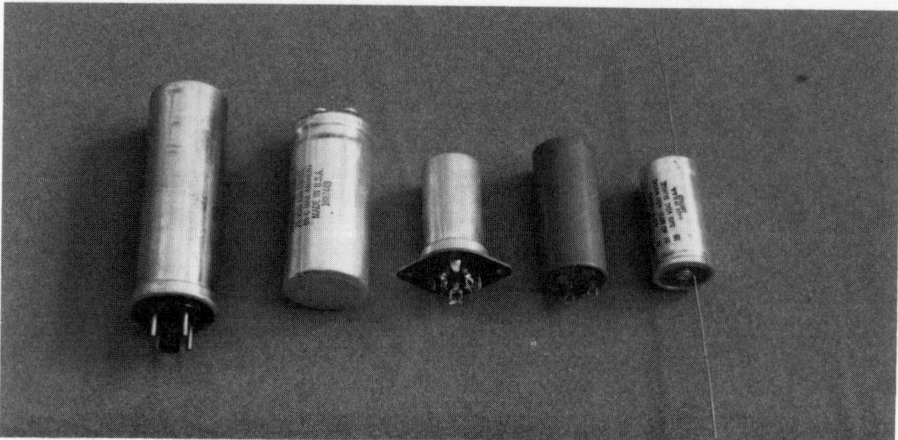

FIGURE 13.3 Electrolytic capacitors

13.4 *CAPACITORS IN SERIES*

When capacitors are placed in series, the total capacitance of the group is reduced to a value less than the smallest. The relationship is expressed by a formula as follows:

$$\frac{1}{C_T} = \frac{1}{C_1} + \frac{1}{C_2} + \frac{1}{C_3} + \cdots + \frac{1}{C_n}$$

To understand this formula, consider two equal capacitors in series as shown in Figure 13.5.

When the switch is closed, electrons will rush to plate *a* where they will repel an equal number of electrons from plate *b*. These electrons have no place to go except plate *c* of C_2 where they will repel

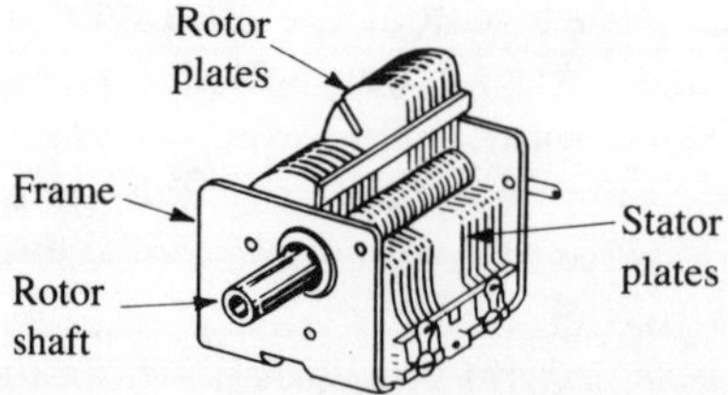

FIGURE 13.4 Typical variable capacitor with rotor plates half engaged

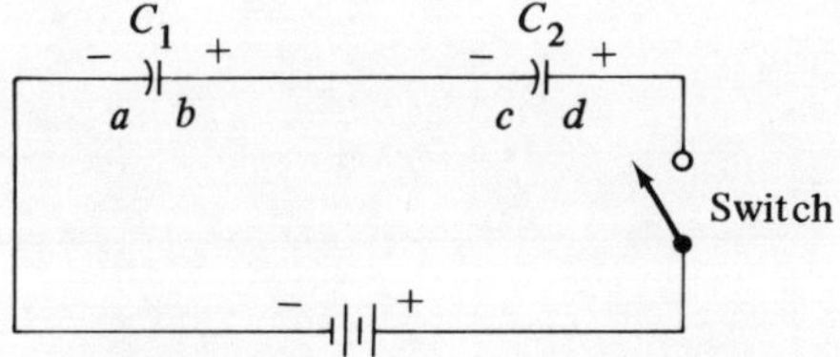

FIGURE 13.5 Two equal capacitors in series

an equal number of electrons from plate *d* which go to the positive terminal of the battery. During the instant when the flow takes place, the
quantity of electrons moved will be such that the voltages across C_1 and
C_2 together equal the total supply voltage *E*. Since the two capacitors are
equal, the voltage across each will be half of *E**. If only one of the capacitors had been in the circuit, enough electrons would have moved to
bring the voltage across the capacitor to an amount equal and opposite to
E. Therefore, twice as many electrons would have moved from the negative terminal of the battery and twice as many would have entered the
positive terminal. Obviously, then, the two equal capacitors cut the capacity in half.

Unequal capacitors cause a similar effect. However, the
voltage will be highest across the smallest capacitor (where the electrons
are the most crowded) and lowest across the largest capacitor.

As indicated by the formula in Section 13.1, capacitive reactance is inversely proportionate to capacitance. The reactances in series
merely add together to give total capacitive reactance as follows:

$$X_{C_T} = X_{C_1} + X_{C_2} + X_{C_3} + \cdots + X_{C_n}$$

CAPACITORS IN PARALLEL 13.5

In a parallel connection, one is really adding plate areas together. As a
result, the total capacitance is the sum of the capacitances as shown in
the formula:

$$C_T = C_1 + C_2 + C_3 + \cdots + C_n$$

The illustration in Figure 13.6 shows why parallel connection is, in effect, the adding of plate areas.

---

*Kirchhoff's voltage law. See Section 3.5.

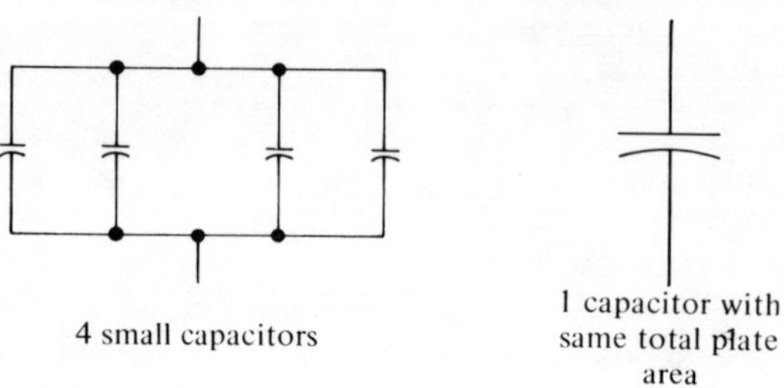

FIGURE 13.6 Capacitors in parallel

Because an increase in capacitance causes a reduction in capacitive reactance, the total reactance follows the reciprocal relationship:

$$\frac{1}{X_{CT}} = \frac{1}{X_{C1}} + \frac{1}{X_{C2}} + \frac{1}{X_{C3}} + \cdots + \frac{1}{X_{Cn}}$$

13.6 *RC TIME CONSTANT*

In the discussions of Figure 13.5 and 13.6, the flow of electrons to the surfaces of the capacitors could be described as instantaneous in an ideal situation where no circuit resistance exists. However, when resistance is also present in the circuit, current cannot exceed the Ohm's law value of E/R. Therefore, a significant amount of time is required for the electron flow to be sufficient to create an equal voltage across the capacitance to oppose the source voltage. In fact, theoretically, the opposing voltage is never completely reached, but for all practical purposes it can be assumed to be at that voltage after about five time constants.

Filling a tire with air is analogous to charging a capacitor. The resistance to the flow of air is provided by the valve between a pressure source and the tire. If the tire is a size suitable for a bicycle, very little time is required to reach a sufficient pressure. On the other hand, if it is an automobile or truck tire, much more time is required to fill it. Similarly, a large-value capacitor requires more time to charge than smaller one, given the same circuit resistance and voltage.

For a circuit consisting of a resistance in series with a capacitance, the time constant is the number of seconds that is equal to the product of the resistance (in ohms) and the capacitance (in farads).

$$\tau_c = RC$$

Consider the *RC* circuit in Figure 13.7. When the switch is closed, current will flow from the battery to charge the capacitor. It will,

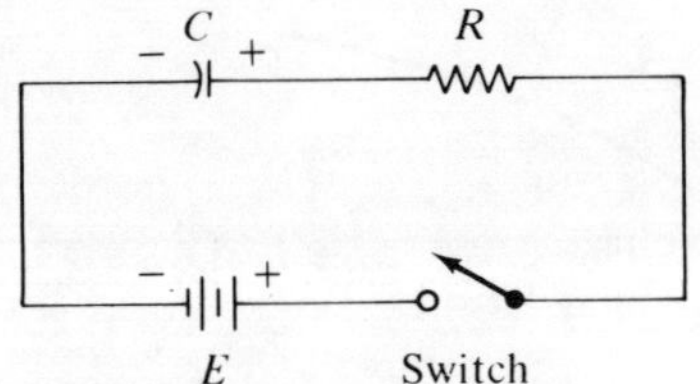

FIGURE 13.7   RC circuit

of course, be limited at the beginning to a maximum value of E/R. Immediately after electrons reach the capacitor, voltage opposing the source voltage begins to build. This causes a reduction in current flow. The graph of current flow is shown in Figure 13.8. After one time constant the current will have been reduced by 63.2 percent to a value of 36.8 percent of the E/R value. In the second time constant the reduction will be 63.2 percent of the remaining value, and so on.

As an illustration, suppose the capacitance C is 10 microfarads, and the resistance is 2 megohms with a source voltage of 100 volts. The time constant is then

$$\tau_c = RC = (2 \times 10^6)(10 \times 10^{-6}) = 20 \text{ seconds}$$

When the switch is closed, the instantaneous current is

$$\frac{E}{R} = \frac{100}{2,000,000} = 50 \ \mu A$$

After 20 seconds, this current would be reduced to 36.8 percent or 18.4 microamperes. At the end of 40 seconds the current would be reduced to 36.8 percent of the 18.4 microamperes (6.8 microamperes).

Once the capacitor is charged, if the source voltage is removed and replaced by a conductor, the capacitor will discharge through the resistor. Again the RC time constant prevails, and the graph of current follows the same pattern as in Figure 13.8, but the direction of current flow is the reverse of that which originally charged the capacitor.

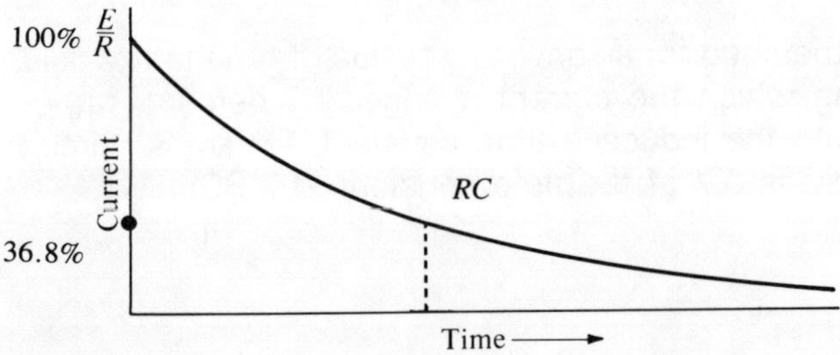

FIGURE 13.8   Effect of capacitor on current

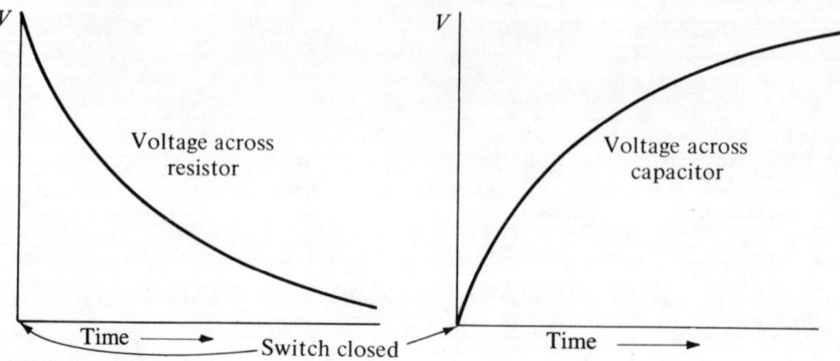

FIGURE 13.9   Voltage change

Discharge of a capacitor can be visualized as similar to a filled water bucket emptying through a leak in its bottom. The hole causing the leakage is like a resistance in series with the bucket (the capacitor). At first the water (charge) will come out in a fast stream, but as time goes by the stream weakens and eventually becomes a mere drip.

In any closed circuit the supply voltage is always equal to the sum of the voltage drops in the rest of the circuit. In the circuit illustrated in Figure 13.7, at the moment the switch is closed, current flow will be maximum through the resistor as the capacitor begins to charge. At that instant the voltage across the resistor will be equal and opposite to the battery **EMF.** Moments later the capacitor will have received some charge and the current through the resistor will have decreased so that the sum of the capacitor and resistor voltage drops will be equal to the battery voltage. A graphic representation of the changes in voltage is shown in Figure 13.9. Naturally, the voltage changes follow the same pattern of change with time as for current.

13.7 *PRACTICAL TIME CONSTANT*

Section 12.10 explained the need for a convenient value of time that would indicate the period during which the current changed 50 percent, rather than 63.2 percent. As with the inductive time constant, for an *RC* circuit the *practical* time constant is 0.7 of the time constant (0.7 RC).

CAPACITOR TIME CONSTANT

Equipment Needed

    1—VTVM
    1—0.68-microfarad capacitor
    1—Watch or clock with sweep-second hand
    1—Breadboard
    1—30-volt power supply

PROCEDURE

1. Mount the 0.68-μf capacitor on the breadboard and connect the VTVM across it.
2. Connect the power supply ground to the same terminal as the ground of the VTVM.
3. Turn the power supply on and set it to 24 volts.
4. While watching the second hand of the clock or watch, momentarily touch the positive probe of the power supply to the positive terminal of the VTVM.
5. Note the voltage on the VTVM at the end of 3 seconds.
6. Successively record the VTVM reading at 6, 10, 15, and 20 seconds.
7. Plot your data on the grid in Figure A. Draw a smooth curve that best satisfies the observed values.
8. The VTVM probably has an input impedence of 11 megohms. Calculate the time constant and the practical time constant of the resistance and the capacitor.
9. Compare your graphic results with the calculations just made.

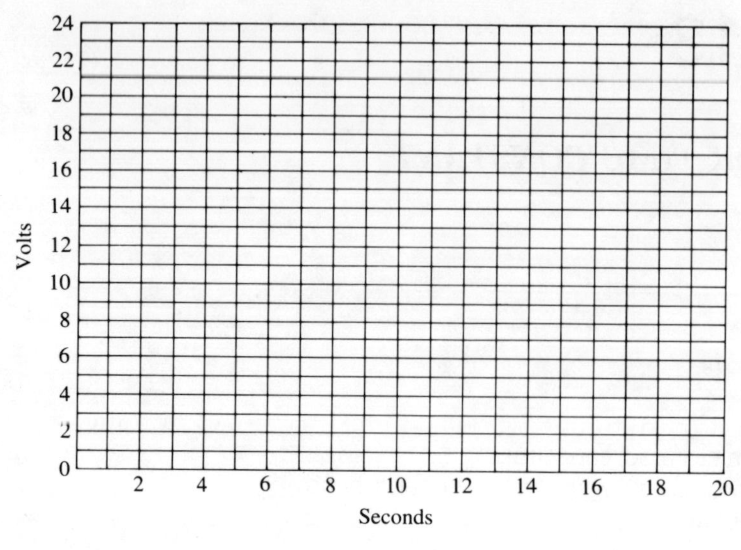

FIGURE A   Time constant graph

QUESTIONS

Select a number, formula, word, or words from the following list to fill each of the blanks in the questions.

alternating	microfarad	second	3.6
area	microhenry	separation	10
current	microsecond	stator	29
dielectric	millifarad	strength	42
direct	millihenry	thickness	60
electrolyte	millisecond	voltage	122
electrolytic	nanosecond	$1/(2\pi fC)$	132
farad	picofarad	$2\pi fC$	550
frequency			

1. The insulating material in a capacitor is called the __________ and the number of times it increases the capacitance over what a vacuum would provide is called the __________ constant.

2. Capacitance can be increased by enlarging plate __________ or decreasing plate __________ .

3. The number of volts per unit of thickness that an insulator can withstand without rupturing is called the __________ __________ .

4. The capacitive reactance caused by a sine wave can be calculated from the frequency and capacitance value with the formula: $X_c =$ __________ .

5. The stationary portion of an air-dielectric variable capacitor is called the __________ .

6. In a series circuit of two capacitors of 220 microfarads and 330 microfarads respectively, the total capacitance is __________ __________ .

7. Parallel capacitors of 47 microfarads and 75 microfarads have a total capacitance of __________ __________ .

8. If a 15-microfarad capacitor is placed in series with a 240-K resistor, the time constant is __________ .

9. If a 250-microfarad capacitor is placed in series with a 240-K resistor, the time constant is __________ .

10. If the capacitor in Question 9 is charged to 50 volts, it will require __________ seconds to discharge it through the 240-K resistor to 25 volts.

11. Capacitance is measured in a unit called the __________ . A capacitor that would store one coulomb of charge at one volt would have a value of one __________ .

12. If a capacitor will store 0.001 coulomb of charge at 100 volts, its capacitance is __________ .

13. Excessive current will flow and destroy an __________ capacitor if a voltage of improper polarity is applied.

14. A capacitor opposes a change in __________ .

15. A capacitor is an open circuit to __________ current.

14

Phase Effects, Impedance, Filters

As long as electrical problems are confined to direct current or to purely resistive circuits the solutions remain relatively simple. Resistances, currents, or voltages can be added together, multiplied, or divided easily, and at worst, the use of reciprocals may first be required.

As soon as alternating current is introduced, however, many complications arise. Our considerations of inductors and capacitors have shown that changing current causes some peculiar effects that may make calculations a bit sticky.

Probably the most troublesome peculiarity is the shifting of **phase** (the time relationship of the zero and maximum points of two or more waves).

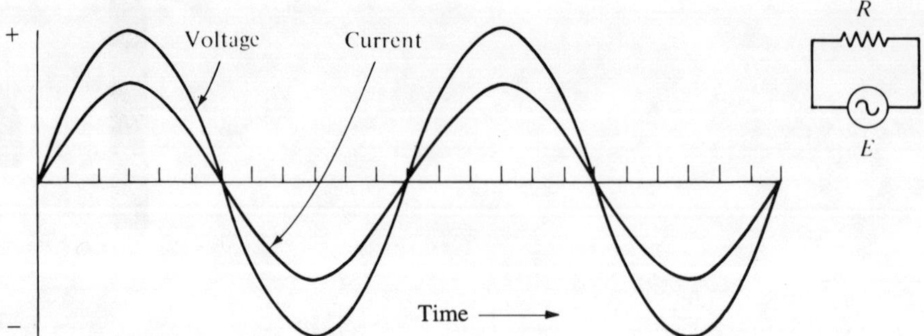

FIGURE 14.1   Illustration that current in a resistor occurs in phase with the voltage

14.2 *PHASE IN A RESISTANCE CIRCUIT*

When an alternating voltage is applied across a resistance, current flow occurs in direct proportion to the changing voltage. In Figure 14.1 the effect is illustrated. The fact of the current rising and falling at the same time as the voltage is referred to as being **in phase.**

14.3 *PHASE IN AN INDUCTIVE CIRCUIT*

When an alternating voltage is applied to an inductor there is a time delay after the voltage reaches a maximum before the current reaches its maximum. This is due to the inductive reactance previously described. The effect is illustrated in Figure 14.2. Since the current in the illustration reaches a maximum later than the voltage, the current is said to **lag** the voltage.

14.4 *PHASE IN A CAPACITIVE CIRCUIT*

When a capacitor is placed across a voltage source, current is maximum when the voltage across the capacitor is minimum. Thus, current in a capacitor is said to **lead** the voltage. The waveforms are illustrated in Figure 14.3.

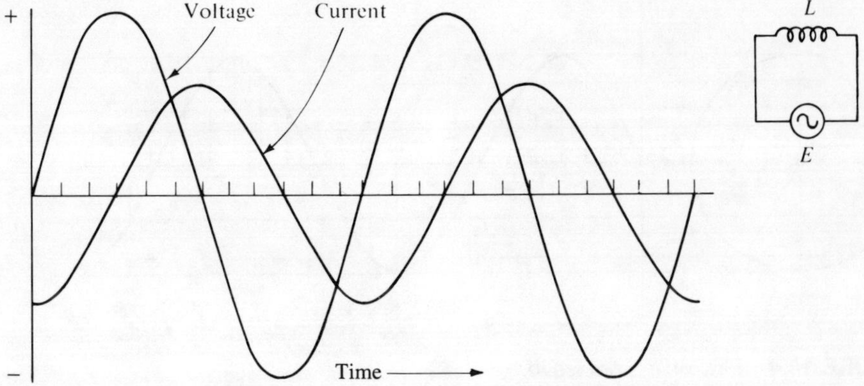

FIGURE 14.2   Illustration that current in an inductor lags voltage

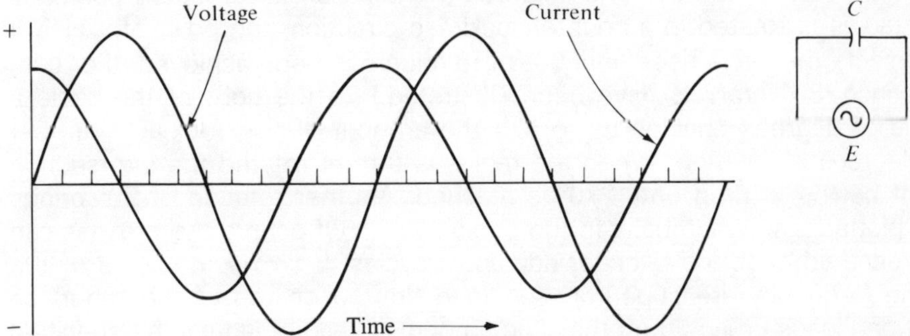

FIGURE 14.3   Illustration that current leads voltage in a capacitor

MEASUREMENT OF PHASE 14.5

In each of the examples so far shown, phase has been indicated in a relative way. That is, current and voltage were in phase, or one was either leading or lagging the other. No *quantitative* measure was specified for the *amount* of lead or lag. This measurement could be done in terms of time except for the fact that it would then be necessary to state the frequency being discussed. Fortunately, there is an easy solution to this problem. By far the most common type of wave used in electronics is a sine wave since this wave is what results from the oscillations of a tank circuit or from the output of a power alternator. A sine wave can be derived by projecting the rotation of a radius around a circle. This has been done in

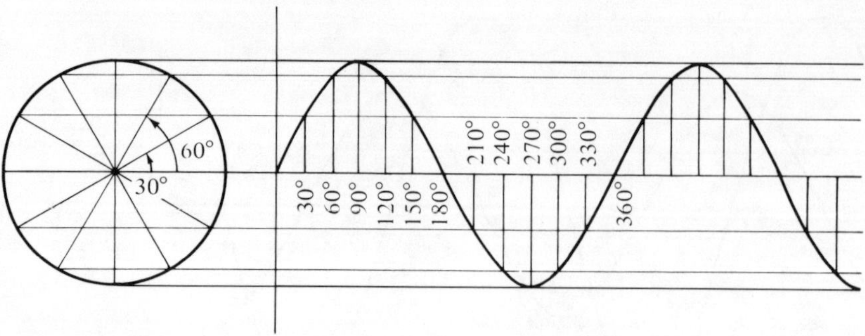

FIGURE 14.4   Plot of a sine wave

Figure 14.4. The position of the radius when it lies horizontally reaching to the right side of the circle is taken as the base or zero rotation point. As the radius is rotated in a counterclockwise direction, the point at which it intersects the circle rises until 90° of rotation has been achieved. (For convenience, 30°-intervals have been illustrated.) To the right of the circle a scale of degrees enables us to plot the amount of rise for each number of degrees of rotation. When the radius is further rotated the intersection point begins to drop until 270° is reached. Another rotation of 90° brings the plotted sine wave back to the original level. Of course the process can be repeated over and over for additional cycles. If two sine waves of the same frequency are superimposed (one drafted on top of the other) so that one leads or lags the other, the amount of lead or lag can be specified in degrees.

      In the case of a pure inductive circuit as shown in Figure 14.2, it can be seen that the current in a pure inductive circuit lags the voltage by 90°. Also, it should be noted that in a pure capacitive circuit, the current leads the voltage by 90°.

14.6 *PHASE IN AN RL CIRCUIT*

If a resistor is placed in series with an inductor, the effect of inductive reactance must be added to the resistance to determine the total impedance. However, recall that the current in the inductor will be out of phase with the voltage. Thus, since the current in the two components must be the same, the voltage will reach maximum across the inductor 90° ahead of the voltage across the resistor. When the voltages at each instant across the resistor and across the inductor are added, it can be seen that they

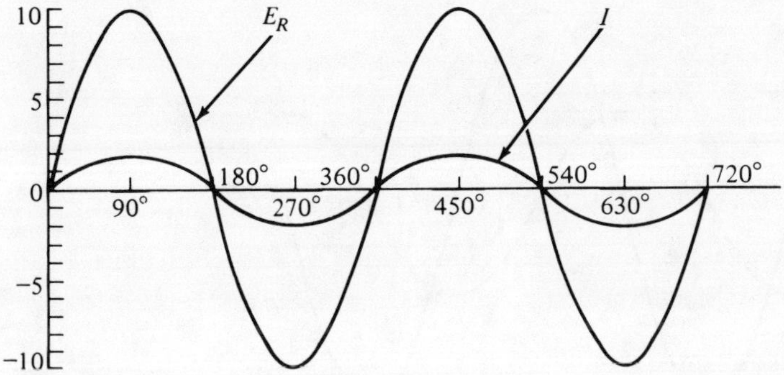

FIGURE 14.5 Voltage and current for resistor

total to the impressed voltage at that instant. However, their peak voltages do not occur together so the *total* voltage *peak* is less than the sum of their peaks Likewise, the rms voltages cannot be added directly. To get a clear idea of the problem, consider a series circuit where a current with a peak value of 20 mA flows through a 500-ohm resistance and a 250-ohm inductance. The peak voltage across the resistor will thon bc

$$(20 \times 10^{-3})(5 \times 10^2) = 10 \text{ volts}$$

The graph of current and voltage is as shown in Figure 14.5. The peak voltage across the inductor will be

$$(20 \times 10^{-3})(2.5 \times 10^2) = 5 \text{ volts}$$

Of course, this curve will lead the current by 90°, so when it is plotted on the same graph as in Figure 14.5, the result is Figure 14.6.

Total voltage at other instants can be found algebraically by adding the instantaneous voltages across the two components. For in-

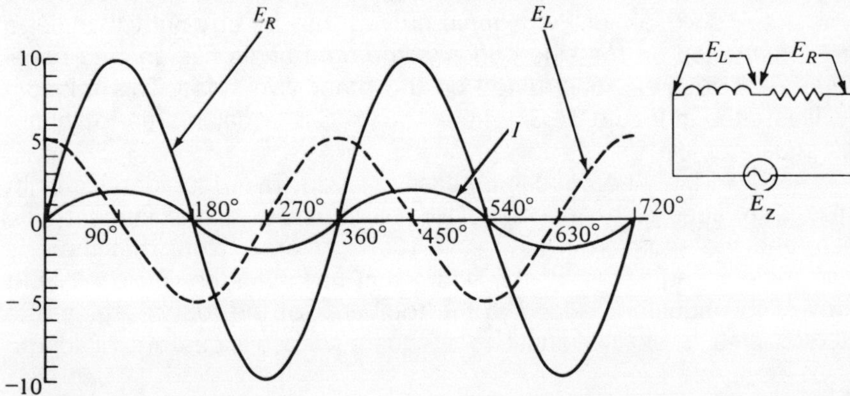

FIGURE 14.6 Voltages across resistance and inductance in relation to current

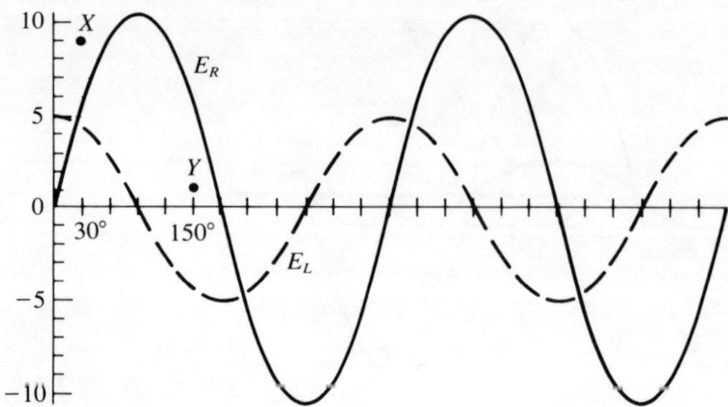

FIGURE 14.7   Points established by sum of simultaneous instantaneous voltages

stance, at 30° the resistor voltage is 5 V, and the inductance voltage is approximately 4.3 V. Their total is 9.3 V, and has been plotted as point *X* on Figure 14.7 to represent a point on the curve for total voltage. Similarly, at 150° the resistor voltage is 5 V, but the inductor voltage is −4.3 V. Now their total is 0.7 V. A point *Y* identifies this sum in Figure 14.7. If this process is repeated for many points and a heavy line is drawn through them, the curve of total voltage (E_2) is as shown in Figure 14.8.

Fortunately one does not need to go through this laborious process. There is an easier way involving a principle known as the *Pythagorean theorem*.

14.7 *PYTHAGOREAN THEOREM*

Although it had been known for well over 1000 years, about 2600 years ago the Greek philosopher Pythagoras proved that for any right triangle, a square constructed on the side opposite the right angle has an area equal to the sum of squares constructed on the other two sides. This relationship is illustrated in Figure 14.9, where square *C* is equal in area to square *A* plus square *B*.

This theorem can be easily proven. Take four identical right triangles with sides *a*, *b*, and *c* and form a square with the longest side of each, and the situation in Figure 14.10a is created. Note that there is a square hole in the center of the figure that has sides equal to *b* − *a*. If the area of this square is added to the total area of the four triangles, the grand total area is exactly equal to a square whose sides are of length $c(c^2)$.

If the four triangles and added square with sides equal to *b* − *a* are now rearranged, Figure 14.10b can be constructed. Its area must be the same as before because it is made of the same components. No-

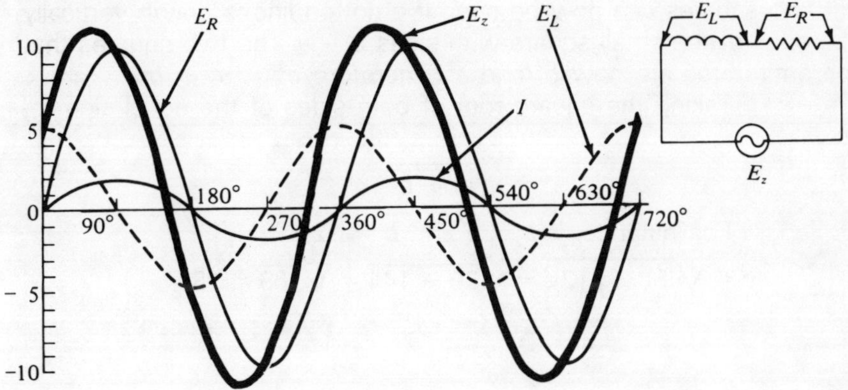

FIGURE 14.8 Total voltage drop across resistance and inductance

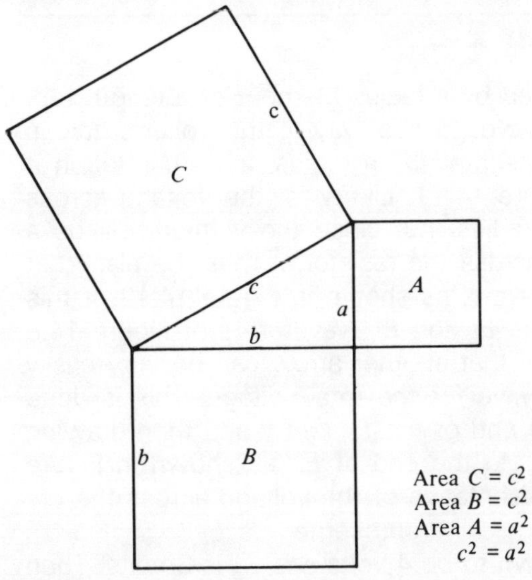

Area $C = c^2$
Area $B = c^2$
Area $A = a^2$
$c^2 = a^2 + b^2$

FIGURE 14.9 Pythagorean theorem

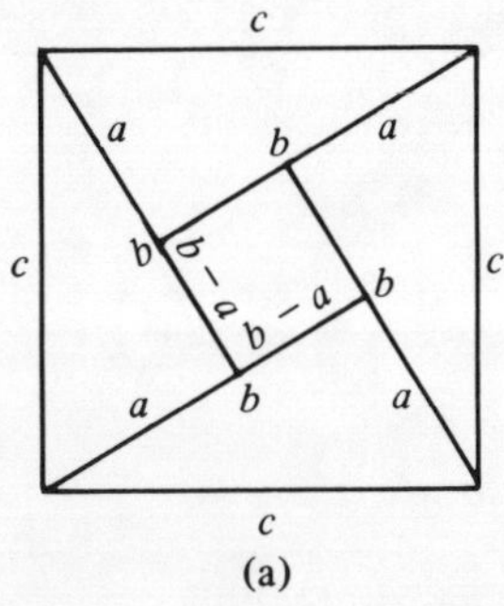

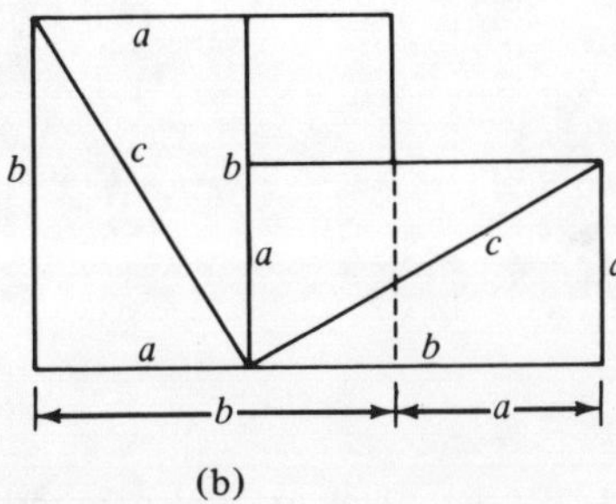

FIGURE 14.10 Pythagorean theorem proof

tice that two squares can now be seen if a dotted line is drawn vertically from the edge of the small square with sides $b - a$. The two squares that total the entire area are now b^2 and a^2. Therefore $c^2 = a^2 + b^2$.

Taking the square root of both sides of the equal sign results in

$$c = \sqrt{a^2 + b^2}$$

For instance, if $a = 5$ and $b = 12$,

$$c = \sqrt{(5)^2 + (12)^2} = \sqrt{25 + 144} = \sqrt{169} = 13$$

14.8 *CALCULATING THE TOTAL VOLTAGE*

Every sine wave can be represented by a radius element of a length representing the peak value of the wave. In this way, if the voltage across the resistor in Figure 14.8 is represented by a radius, at zero rotation it would appear as the arrow in Figure 14.11. Likewise, the voltage across the inductor can be shown as a radius, but since the voltage will be a maximum 90° before the voltage across the resistor, it must be placed at an angle of 90° with the resistor arrow, as shown in Figure 14.12. If this combination is rotated together, the E_L and E_R waveforms of Figure 14.8 are produced. The amazing thing is that another arrow can be very easily found that will develop the sine wave for the total voltage. This is done by moving the E_L arrow out to the end of the E_R arrow and then drawing an arrow from the beginning of E_R to the end of E_L, as shown in Figure 14.13. This resulting arrow is representative of the voltage across the entire impedance and is, therefore, logically represented as E_Z.

Suppose E_L is known to be 4 volts and E_R is 3 volts. Then $E_Z = \sqrt{(4)^2 + (3)^2} = \sqrt{16 + 9} = \sqrt{25} = 5$ volts. Of course, E_Z can also be found by using trigonometry.

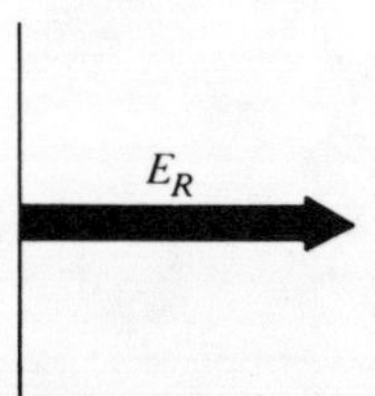

FIGURE 14.11   Resistor voltage represented at 0° with respect to current

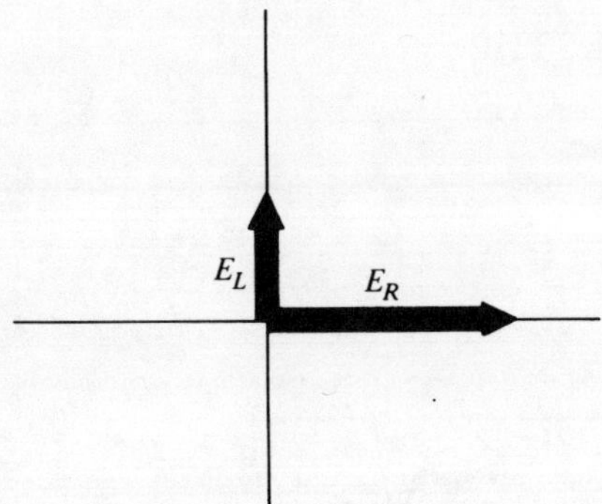

FIGURE 14.12 Inductor voltage added to illustrate a lead of 90° with respect to current

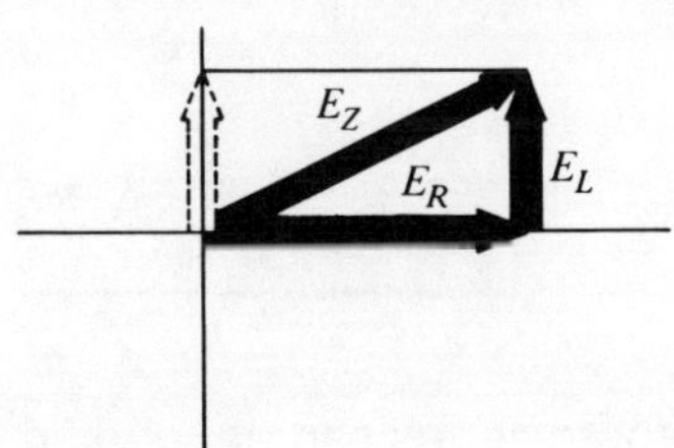

FIGURE 14.13 Representation of voltage across impedance

Impedance is calculated from the Ohm's law relationship previously stated as

$$I = \frac{E}{Z}$$

$$E = IZ$$

$$Z = \frac{E}{I}$$

The impedance of the circuit just described is

$$Z = \frac{E}{I} = \frac{5}{I}$$

For illustration, assume that $I = 3$ amperes. Then, $Z = 5/3 = 1.67$ ohms. The same result is obtained if the voltages across the components are used to determine their individual impedances. For the inductor, the reactance is

$$X_L = \frac{E_L}{I} = \frac{4}{3} = 1.33 \ \Omega$$

For the resistor the resistance is calculated

$$R = \frac{E_R}{I} = \frac{3}{3} = 1 \ \Omega$$

The total impedance is, then,

$$Z = \sqrt{(X_L)^2 + (R)^2} = \sqrt{\left(\frac{4}{3}\right)^2 + (1)^2} = \sqrt{\frac{16}{9} + 1}$$

$$= \sqrt{\frac{16 + 9}{9}} = \sqrt{\frac{25}{9}} = \frac{5}{3} = 1.67 \ \Omega$$

14.9 *PHASE IN AN RC CIRCUIT*

The phase relationships of a resistor and capacitor in series can be worked out in essentially the same manner as was explained for a resistor and inductor.

In this case the current in the circuit leads the voltage across the capacitor by 90°. This effect can be represented as shown in Figure 14.14. A single arrow (**vector**) can be found to show the total circuit voltage by the method previously illustrated, as in Figure 14.15. As before, the total voltage can be calculated as

$$E_Z = \sqrt{(E_C)^2 + (E_R)^2}$$

For instance, if the capacitor voltage is 6 volts and the resistor voltage is 8 volts,

$$E_Z = \sqrt{(6)^2 + (8)^2} = \sqrt{36 + 64} = \sqrt{100} = 10 \ \text{V}$$

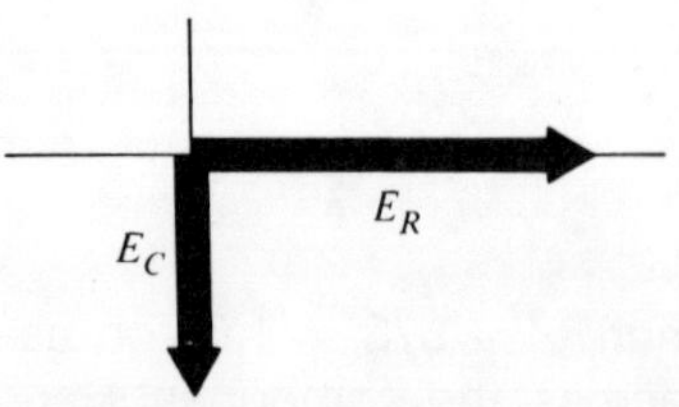

FIGURE 14.14   Phase relationship of the voltage across a resistor and capacitor in series

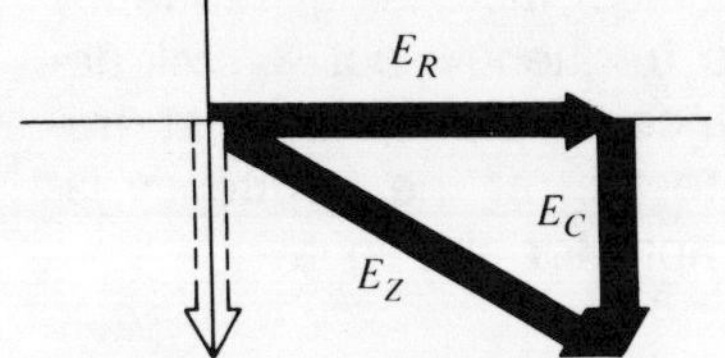

FIGURE 14.15   Resultant voltage across a resistor and capacitor in series

Similarly, the method of calculating impedance in the series circuit is

$$Z = \sqrt{(X_C)^2 + R^2}$$

If the current in the circuit is 2 amperes,

$$X_C = \frac{E_C}{I} = \frac{6}{2} = 3 \ \Omega$$

$$R = \frac{E_R}{I} = \frac{8}{2} = 4 \ \Omega$$

$$Z = \sqrt{(3)^2 + (4)^2} = \sqrt{9 + 16} = \sqrt{25} = 5 \ \Omega$$

PHASE IN AN LCR CIRCUIT 14.10

Naturally, many cases exist where L, C, and R are in series. Since the voltages across the capacitor and inductor are 180° apart, they tend to cancel one another when they are added (as in Figure 14.16). Thus, the total voltage can be found as follows:

$$E_Z = \sqrt{(E_L - E_C)^2 + E_R^2}$$

Similarly, impedance can be found by the formula,

$$Z = \sqrt{(X_L - X_C)^2 + R^2}$$

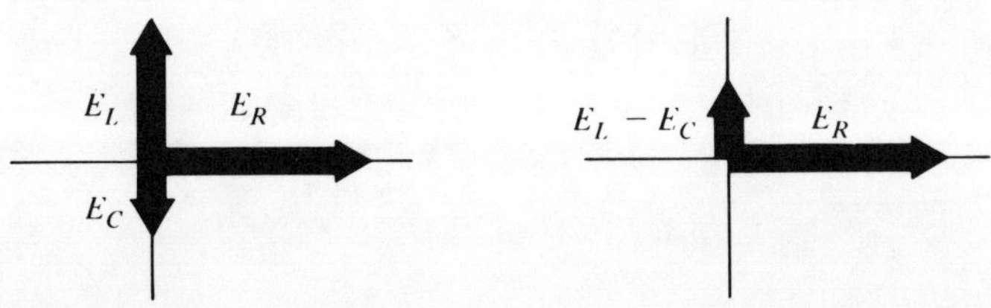

FIGURE 14.16   Result of vector addition of E_L and E_C in a series circuit

It is interesting to note the effect of increasing frequency in this type of circuit. X_L will increase with the frequency, but X_C will decrease ($X_L = 2\pi fL$; $X_C = 1/2\pi fC$). If reactance is plotted against frequency, the graph in Figure 14.17 results. Note that at the frequency labelled f_0, the reactances of the inductor and capacitor are equal.

14.11 *SERIES RESONANCE*

From the foregoing it is natural to question what happens when X_C and X_L are equal in an *LCR* series circuit. From the impedance formula it is apparent that Z would be equal to R. This is, in fact, what happens. Current will be limited only by the resistance in the circuit. If, for instance, 10 volts is applied across a circuit where $X_L = 100\ \Omega$, $X_C = 100\ \Omega$ and $R = 5\ \Omega$, the current will be

$$I = \frac{E}{R} = \frac{10}{5} = 2\ \text{A}$$

The voltage across the resistor will be

$$E_R = IR = 2(5) = 10\ \text{V}$$

However, voltage across the inductor will be

$$E_L = IX_L = 2(100) = 200\ \text{V}$$

Across the capacitor it will also be

$$E_C = IX_C = 2(100) = 200\ \text{V}$$

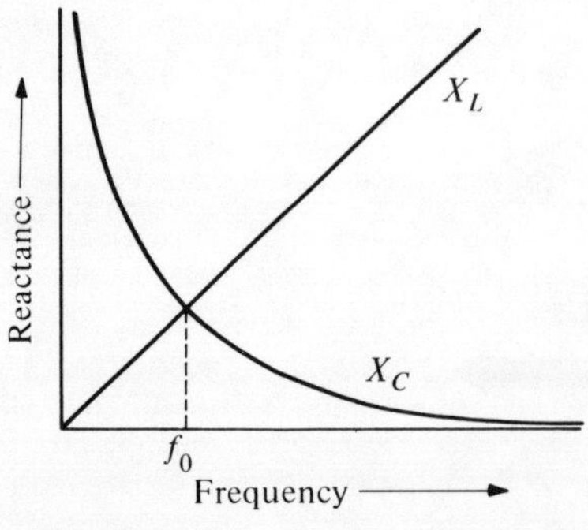

FIGURE 14.17   Effect of frequency increase in an LC circuit

This seems incredible but it is true. The experimenter is warned that extremely high voltages can exist with this condition called **resonance** when only a small voltage is applied. Such voltages are capable of giving the unwary an unpleasant thrill.

In Figure 14.17 the resonant frequency f_0 is illustrated. This value can be calculated for any *LCR* circuit by the following formula:

$$f_0 = \frac{1}{2\pi\sqrt{LC}}$$

IMPEDANCES IN SERIES **14.12**

As discussed earlier, when impedances are placed in series, the current through each component is exactly the same at any instant. However, with an AC sine wave, the voltage across an inductor reaches a maximum 90° before the voltage across a series resistor and 180° before a series capacitor. The voltage across the capacitor lags 90° behind the resistor voltage.

IMPEDANCES IN PARALLEL **14.13**

When a parallel connection is made, as in Figure 14.18, the voltage across each component is identical at each instant. Thus, the current will reach a maximum in the capacitor 90° before the maximum is reached in the

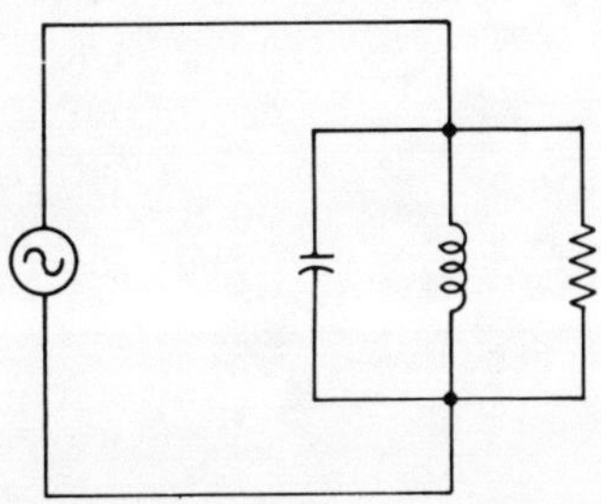

FIGURE 14.18   Impedances in parallel

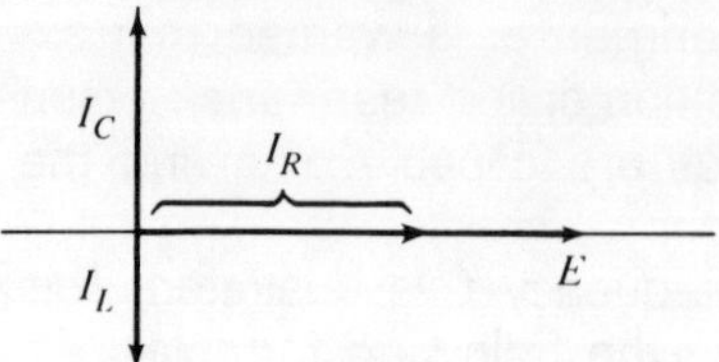

FIGURE 14.19   Relationship of currents in parallel impedances

resistor and 180° before the maximum in the inductor. This can be pictured in vector form, as shown in Figure 14.19.

If the reactance of the inductor and capacitor are equal (resonance), the current flowing in the inductor is in the opposite direction to that in the capacitor so that the net flow is only the current through the resistor. Therefore, if the resistance is high, the impedance of such a resonant circuit is also high. From a practical standpoint, a parallel circuit of a pure capacitance, pure resistance, and pure inductance does not occur. It is nearly impossible to make an inductor that does not contain a significant amount of resistance. Therefore, a more common circuit is one like that shown in Figure 14.20. Of course, the voltage across each leg of the parallel circuit is still the same. However, there is less than 180° difference in phase between the two legs because the voltage across the *LR* branch must now be the vector total of the voltage drops across the inductance and resistance. These two are 90° apart as explained in Section 14.8. The fact of a phase difference between the two legs may be more easily understood if one considers the relationship of current to voltage if the *LR* branch were entirely resistive. In that case, the current would be in phase with the voltage and would lag the current in the capacitive branch by 90°. On the other hand, if the branch were entirely inductive, the current in it would lag the voltage by 90° and the current in the capacitive branch by 180°. Since it is neither of these extremes, the current in the

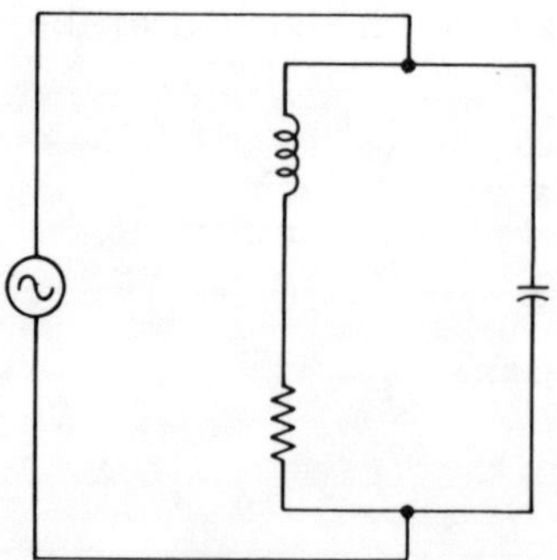

FIGURE 14.20   A capacitor in parallel with an inductor containing significant resistance

inductive branch will lag the voltage by something less than 90° depending upon the amount of resistance. Suppose, for example, that the applied voltage is 50 volts and X_L and X_C are each equal to 500 ohms, but that the internal resistance of $X_L = 200\ \Omega$. The impedance of the inductive branch of the parallel circuit would be

$$Z_L = \sqrt{(500)^2 + (200)^2} = \sqrt{250,000 + 40,000} = \sqrt{290,000} = 539\ \Omega$$

As a result

$$I_C = \frac{50}{500} = 100\ \text{mA}$$

$$I_L = \frac{50}{539} = 93\ \text{mA}$$

The voltage drops across the resistance and inductive reactance are, therefore,

$$E_R = (93 \times 10^{-3})(2 \times 10^2) = 18.6\ \text{V}$$
$$E_L = (93 \times 10^{-3})(5 \times 10^2) = 46.5\ \text{V}$$

If these are plotted, Figure 14.21 is the result.

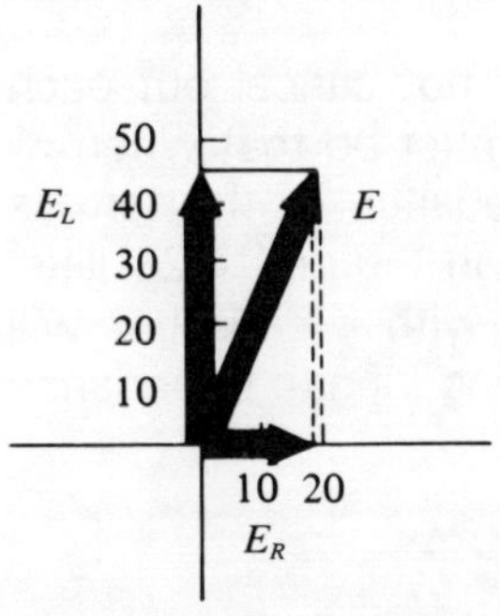

FIGURE 14.21   Phase relationship of inductive reactance voltage drop and applied voltage

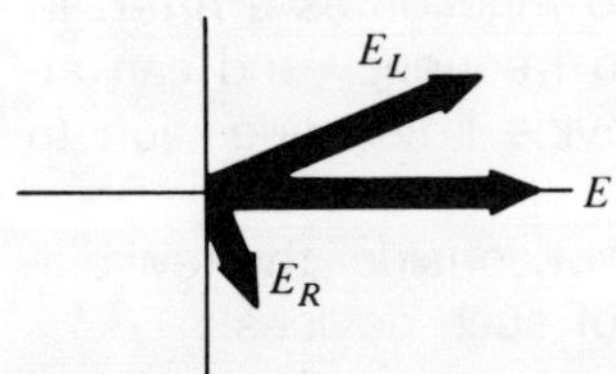

FIGURE 14.22   Rotation of Figure 14.21 to place total voltage in reference position

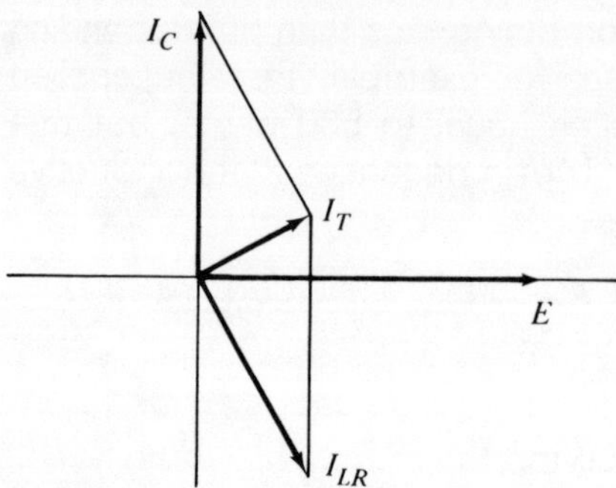

FIGURE 14.23 Total current (I_T) resulting from resistance in an inductor in a parallel tank circuit

This figure can now be rotated so that E is in the horizontal reference position. (This is desirable because in a parallel circuit the voltage across each branch is the same.) The result is Figure 14.22.

If the current in the inductive branch is now plotted so that it lags the inductive voltage by 90°, it will appear as shown in Figure 14.23. In addition, the current in the capacitive branch has been indicated in the figure to lead the applied voltage by 90°. The total circuit current (I_T) that results is also shown.

As the figure indicates, I_C and I_L do not cancel out each other since they are not 180° apart. The conclusion must be that when X_L and X_C are equal in a parallel (tank) circuit the impedance of the tank is lowered by resistance in either branch. The sketch in Figure 14.23 illustrates that any resistance in an inductor in parallel with a capacitor will prevent a tank circuit of equal reactances from reaching infinite impedance at resonance.

14.14 *FILTERS*

One common use of impedance components is to function as a **filter.** In electronics, a filter is a device that is sensitive to frequency and can attenuate signals of unwanted frequencies but provide little opposition to others.

Many arrangements are possible, but consider the few diagrams following that illustrate the potential uses of such devices.

Suppose, for instance, that a circuit is required to pass all frequencies except 1000 Hertz. A series circuit resonant at 1000 Hertz can be connected to ground, as shown in Figure 14.24. The 1000-Hertz signal

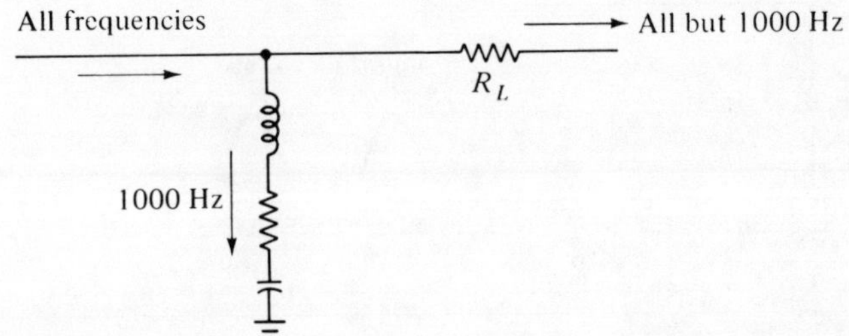

FIGURE 14.24 Frequency trap

thus finds an easy path to ground, but is attenuated through the resistor R_L. All the other frequencies "see" a higher impedance through the inductor/capacitor link to ground and, therefore, go through the load (R_L). This effect can be further enhanced by placing a parallel circuit resonant at 1000 Hertz in series with R_L, as shown in Figure 14.25. In this arrangement, the 1000-Hertz signal sees a very high impedance in the parallel circuit but low impedance to ground. All other frequencies see a high impedance to ground but low impedance through the parallel circuit and the load.

One of the most common uses for a filter is in the construction of a direct-current power supply. An alternating-current voltage is applied to a diode, as shown in Figure 14.26. Without filtering, the result is a pulsating voltage. To modify this to get a steady DC flow, a capacitor and inductor can be added, as in Figure 14.27.

A capacitor will allow current to flow into it as long as the charge across it is less than the applied voltage. As the generator of Figure 14.27 makes the anode of the diode more and more positive, positive charge will flow easily through the diode to the junction of the capacitor and the inductor. The inductor opposes and, therefore, limits the rate of increase in current through it, but the capacitor readily accepts the charge and attracts electrons to its grounded terminal so that the charge across it rises to the generator voltage. As the generator voltage decreases and then changes to a negative polarity, the stored charge on the capacitor supplies

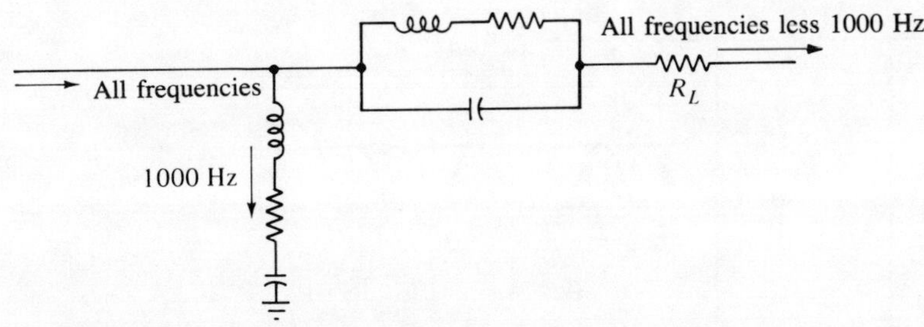

FIGURE 14.25 Improved frequency trap

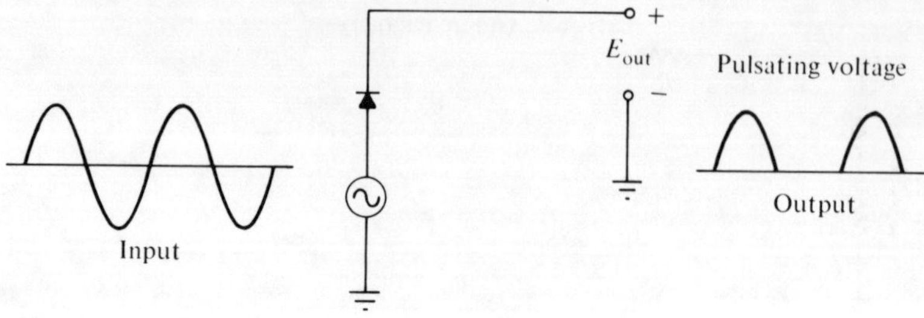

FIGURE 14.26   Simple rectified circuit

current to the inductor. Since any decline in current in the inductor creates a collapsing magnetic field, voltage is induced in the inductor to encourage this flow. The reversal of polarity by the generator has no effect because the diode prevents any flow of negative charges from the anode to the cathode. Observe from the foregoing that the capacitor is the only source of current when the generator is supplying a negative potential to the anode of the diode. The capacitor provides a temporary storage of charge just as a tank stores air from a compressor. The inductor prevents rapid changes in current flow much as an engine flywheel prevents rapid changes in the speed of a machine. The result is a fairly steady flow of direct current. The smoothing of flow can be made even greater by placing another capacitor in the circuit following the inductor as shown in Figure 14.28. The bleeder resistor shown in this circuit is mainly a safety feature that allows the charge of the filter capacitors to "bleed" off when the power supply is turned off, even if it is not connected to a load. It also serves to moderate the percent of change of load in applications where load changes are great.

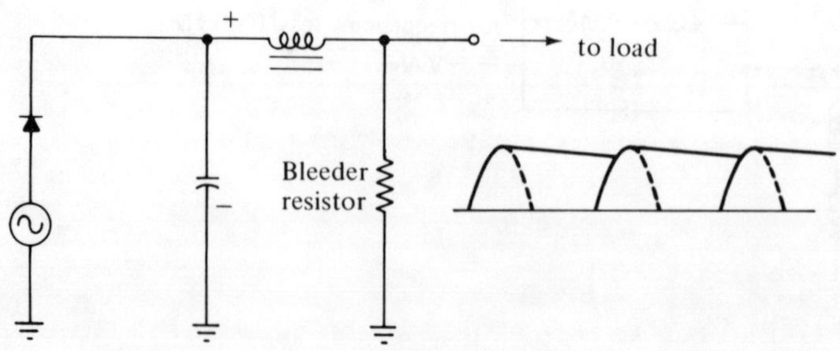

FIGURE 14.27   Rectifier with filter

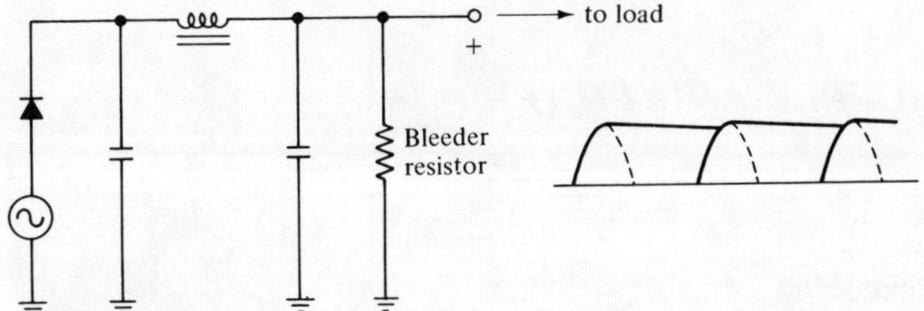

FIGURE 14.28 Rectifier with improved filter (called a π filter)

COUPLING 14.15

Often a capacitor is used to **couple** one circuit to another. This coupling is in reality a sort of filter. Consider, for instance, the operation of a two-stage transistor amplifier as shown in Figure 14.29. Capacitors C_1, C_2, and C_3 permit the alternating signals to be coupled into the first stage, out of it into the second stage, and then out of the second stage of amplification to the output. However, these coupling capacitors block any flow of DC from stage to stage that would interfere with the DC bias each transistor needs to amplify properly. Capacitor C_4 also performs a type of filtering by allowing alternating current to pass from the emitter of Q_2 to ground without much voltage drop. Any DC present must flow through the resistor R_e in parallel with the capacitor. The resulting DC voltage drop makes the emitter more positive than ground. Since no AC flows through R_7, the DC voltage at the base of Q_2 is determined only by the DC drops across R_5 and R_7. Thus, negative feedback due to the AC is minimized and amplification is not degraded.

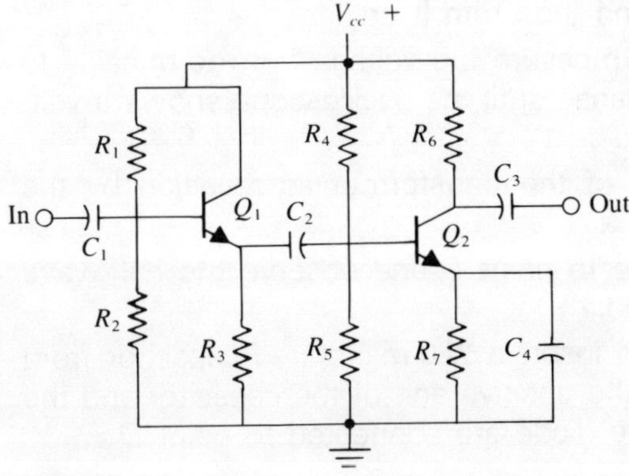

FIGURE 14.29 Circuits coupled by capacitors

POWER SUPPLY AND FILTER

Equipment Needed

    1—Center-tapped transformer
    1—Variac
    1—Oscilloscope
    4—1N4001 Diodes
    2—220-microfarad, 100-V electrolytic capacitors
    1—8-henry choke
    1—10-K, 1/2-W carbon resistor

PROCEDURE

1. Construct the circuit illustrated in Figure A.

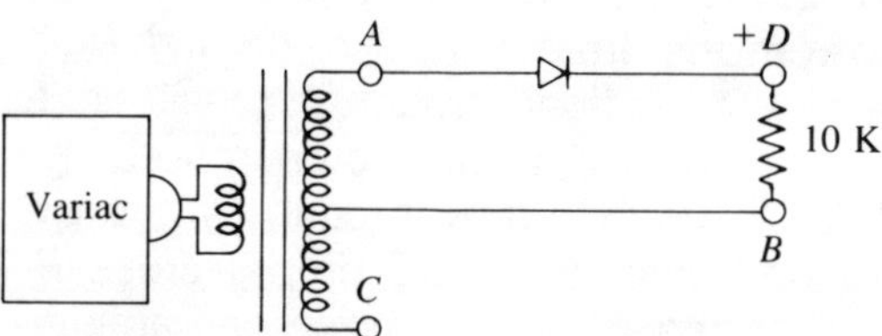

FIGURE A   Half-wave rectifier

2. Set the variac dial to zero and then turn it on.

3. Connect the oscilloscope to measure the voltage from terminal *A* to terminal *B*, then adjust the variac until the oscilloscope shows a voltage of 10 volts peak-to-peak.

4. Note the sine wave output of the transformer as revealed by the oscilloscope.

5. Move the oscilloscope probe to point *D* and observe the half-wave rectification that has occurred.

6. Turn off the variac, and then insert a 220-microfarad capacitor from point *D* to point *B*. Be sure the positive end of the capacitor and the negative end (cathode) of the diode are connected to point *D*.

7. Turn on the variac and note the change in the output. It should be closer to a constant level (DC) than before, but it will still show considerable **ripple** (unfiltered DC pulses) on the oscilloscope screen.

8. Turn off the variac and rebuild the circuit to the arrangement shown in Figure B.

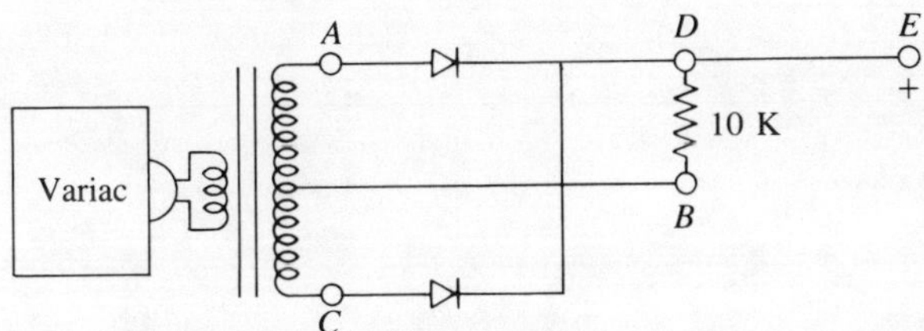

FIGURE B   Full-wave rectifier

9. Connect the ground of the oscilloscope to point B and place the probe on point A (in Figure B).

10. Turn on the variac and adjust it to assure that the voltage is still at 10 volts peak-to-peak.

11. Move the oscilloscope probe to point D and observe the full-wave rectification.

12. Turn off the variac and again insert the capacitor from point B to point D. (Again, observe proper polarities.)

13. Turn on the variac and observe how much less ripple is present than was seen during step 7.

14. Turn off the variac, and then (with the choke between points D and E) connect the second capacitor from B to E. Move the 10-K resistor so that it is in parallel with the second capacitor (from B to E).

15. Move the oscilloscope probe to point E, turn on the variac, and observe the output from E to B. Note that the ripple has almost (if not totally) disappeared. The better smoothing resulted from a π filter.

16. Turn off the variac and rearrange the circuit as indicated in Figure C. This is a configuration called a **bridge rectifier**.

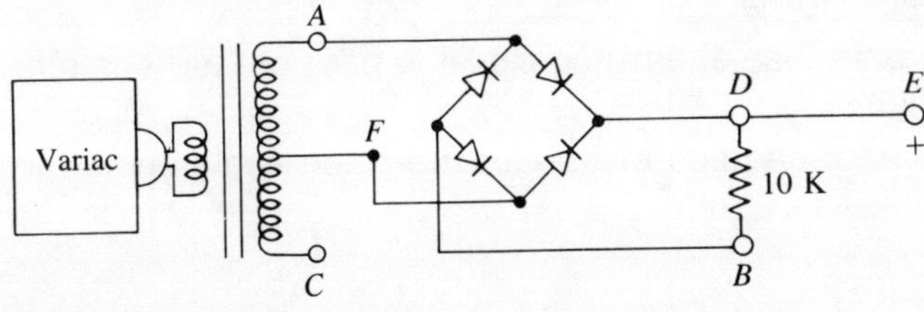

FIGURE C   Bridge rectifier

17. Turn on the variac and adjust as necessary to assure that the voltage from *A* to *F* is still 10 volts peak-to-peak as measured by the 'scope.

18. Move the oscilloscope ground to *B* and the probe to *D* and observe that the output is identical to what you saw in step 11.

19. Repeat steps 12 through 15.

QUESTIONS

Select a number, word, or words from the following list to fill each of the blanks in the questions.

capacitor	leads	volts	13
downward	resistors	10	14
horizontal	resonant	11	90
lags	upward	12	1130

1. Current and voltage are in phase in __________.

2. The voltage across a resistor and a pure inductance in a series circuit are different in phase by __________ degrees.

3. Current __________ voltage in a capacitor.

4. Current __________ voltage in an inductor.

5. Vectors can be used to represent currents and voltages. When used to represent these in a series circuit the current vector is shown as a __________ arrow.

6. In the type of vector diagram described in question 5, a voltage across an inductor would be a vector pointed __________.

7. In a series circuit, if the voltage across the inductance is 20 volts, across the capacitance is 14 volts, and across the resistance is 8 volts, the total circuit voltage is __________.

8. When inductive reactance is equal to capacitive reactance in a series circuit, the circuit is said to be __________.

9. A 0.047-microfarad capacitor, a 0.54-henry inductance, and a 150-ohm resistance are in series across a 50-volt, 1000-Hertz source. The voltage across the capacitor is __________.

10. A __________ will permit an alternating signal to pass but will obstruct a direct-current signal.

11. Plot a sine wave with degrees of rotation from the circle shown in Figure D.

PHASE EFFECTS, IMPEDANCE, FILTERS

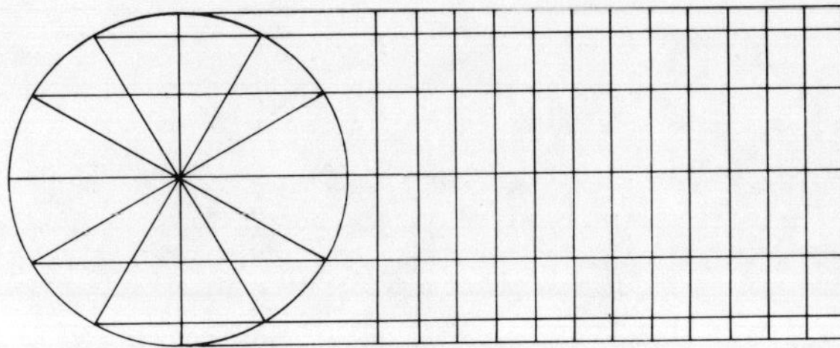

FIGURE D Sine wave graph

12. With a ruler plot on Figure E a vector diagram of a resistive voltage of 10 volts, an inductive voltage of 6 volts, and a capacitive voltage of 1.5 volts in series. Label the resultant total voltage as E, and indicate its scaled value. Use a scale of one volt equal to one-eighth inch. $E =$ ___________.

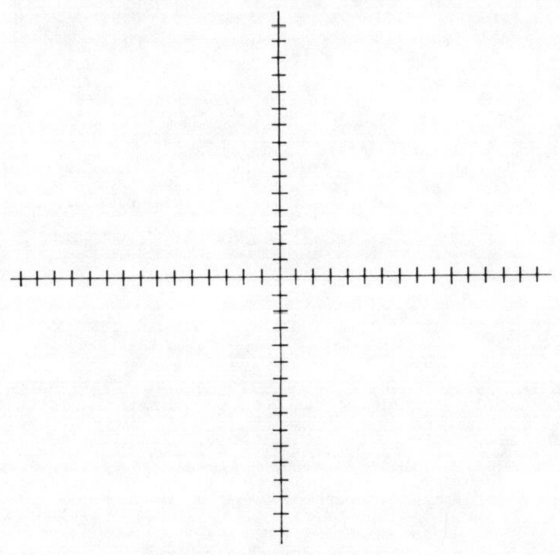

FIGURE E Vector diagram

15

Modern Circuits

The field of electronics has been developing so rapidly that what is modern today is obsolescent tomorrow. The basic facts of electricity do not change, but the hardware of 1985 will be obsolete in 1990. Nowhere is this more evident than in the computer industry. For instance, the microcomputer on which this paragraph is being written was purchased only six months ago, and already new models are available with drastically improved memory capabilities. In the 1970's a computer with all the features of the author's was completely out of the reach of most private individuals' finances.

For many years after Dr. Lee DeForest developed the triode vacuum tube, various kinds of tubes seemed to be the answer to all kinds of control and amplification jobs. A few types of semiconductors were known and used such as the galena crystal (the heart of the famous crystal radio). However, semiconductors have replaced tubes at an accelerating rate since the invention of the transistor in 1948.

It is an interesting fact that the development of commercial television, the invention of the transistor, and the beginnings of the

219

electronic computer all occurred in a short period immediately following World War II. During and following the war, various electronic control mechanisms such as the proximity fuse and missile guidance equipment placed a premium on the space occupied by (and weight of) the electronics package. In addition, the need for an inexpensive method of assembling components into a package that would suffer less failure due to vibration brought about the development of the **printed circuit**.

It was natural that the earliest use of transistors and other semiconductor devices would be on a basis of individual (**discrete**) components. Early transistor circuits copied vacuum tube circuits, which (for several reasons) had always been made this way. For instance, vacuum tubes have a limited life that requires circuitry enabling easy replacement of the tubes. Furthermore, techniques for more compact assembly had not yet been developed.

15.2 *PRINTED CIRCUITS*

A printed circuit (Figure 15.1) can be made from a copper-clad sheet of insulating material coated with a light-sensitive compound that is exposed

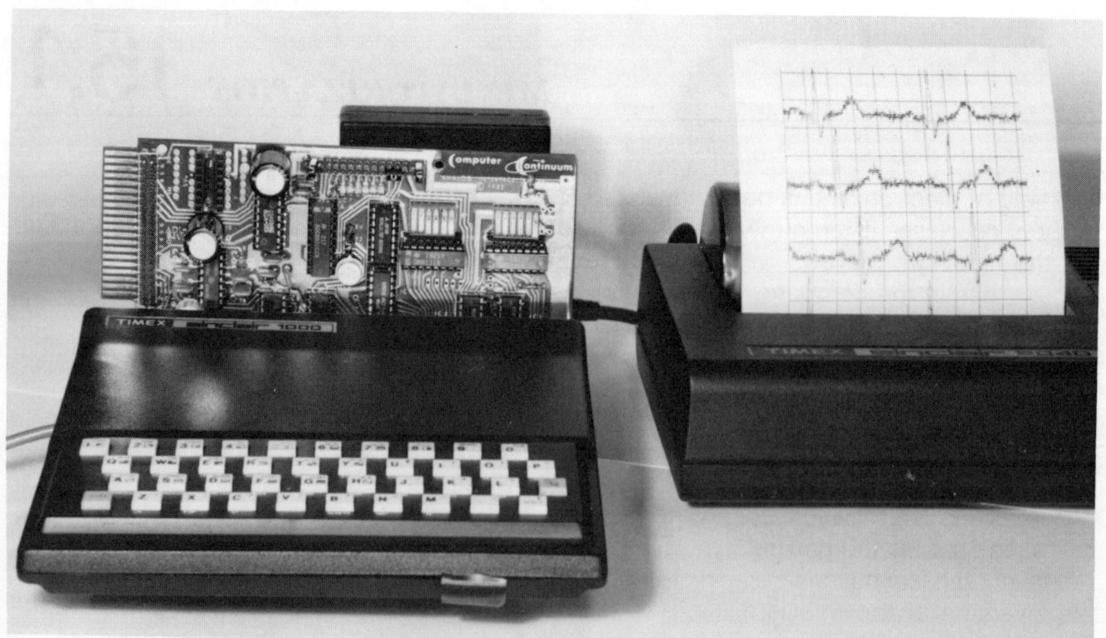

FIGURE 15.1   Printed circuit board (Courtesy of Computer Continuum)

and developed somewhat like a snapshot. The compound (called *photo-resist*) has the quality of resisting the action of a corrosive chemical only where it is exposed to light. A negative with all the necessary circuit connections is used to expose the coated, copper-clad board. The unexposed copper (not protected by the developed image) is then **etched** away with chemicals so as to leave a pattern of copper connections between various points on the insulated sheet. The components are then attached and soldered to make a small package. For production quantities, the photo process just described could be replaced by other means of covering the copper pattern prior to etching. A silk screen process is sometimes used that provides a stencil through which a protective paint can be wiped to form the circuit image.

INTEGRATED CIRCUITS **15.3**

Some **integrated circuits** (IC's) being used today are just a step away from a printed circuit since they provide for the attachment of some discrete components. These are called **hybrid** IC's. Other IC's contain many transistors, diodes, and conductors constructed in a sort of layer-cake fashion on a tiny **chip** of silicon that may be as small as the head of a pin. The circuitry is so shrunken that it can be seen only under a microscope.

In the illustrations shown in Figure 15.2a, two very complex integrated circuits (10,000 transistors each) are shown as they appear repetitively on a disc of silicon prior to being separated and mounted into protective packages. Figure 15.2b shows a similar circuit in greater detail.

IC's are classified as **small-scale integrated** (SSI), **medium-scale integrated** (MSI), **large-scale integrated** (LSI), or **very-large-scale integrated** (VLSI) according to the number of components included on one chip. Modern microcomputers, for instance, use chips containing thousands of transistors, diodes, and resistors in a space about the size of a capital *M* in this book.

IC's are packaged in three common ways. They may be in a circular (transistor) can, in a flat pack, or in a DIP.

IC PACKAGES **15.4**

The **dual in-line package** (DIP) has become very popular, and a few types are illustrated in Figure 15.3. DIP's have an even number of contact pins

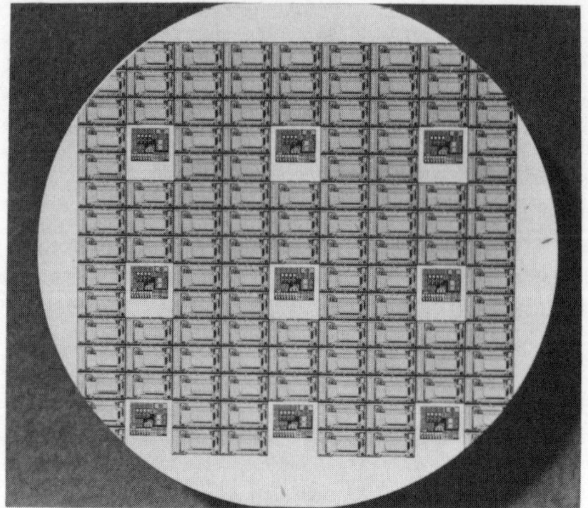

(a) Wafer, approximately 3-inch diameter

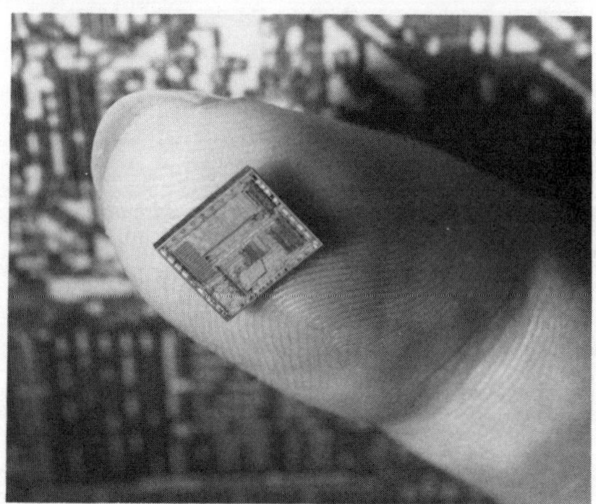

(b) Chip of similarly complex circuit

FIGURE 15.2   Integrated circuits (Courtesy of TRW and Intel Corp.)

that have the same lateral spacing whether there are 8, 14, 16, 24, 36, or even more connections. This standard spacing permits use of assembly boards or IC sockets that will accommodate any size DIP as long as the number of contacts does not exceed the board or socket capacity. Of course, IC's can also be mounted directly onto a printed circuit board.

Many hundreds of IC circuits are now available at moderate cost. In most cases the actual circuit arrangement within the IC is not known to the technician using it, nor is such information needed. Generally, what is most important is type of IC (amplifier, oscillator, etc.), the identity of each pin, and the input or output voltages and signals specified for each pin. Operating voltages are shown on the device specification

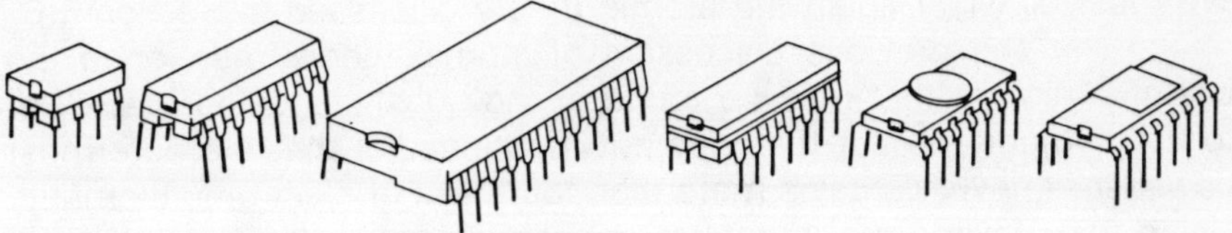

FIGURE 15.3 Dual in-line packages

sheets and should be carefully followed. Pins are counted on DIP's in the following manner: Hold the package with the pins pointing toward you and with the notched end pointed to your left. Pin number 1 will be in the upper left corner. Count clockwise around the pins. The highest-numbered pin will be in the lower left corner.

A **flat pack** is another packaging method used, and it allows the IC to be placed (flat) on a printed-circuit board so the leads will lie directly on circuit connections (the copper side) which are then soldered. Thus, drilling the board is not required and sockets are not used. This setup is economical, but replacing a defective IC is difficult.

Counting pins on the flat pack requires identification first of pin 1 (usually identified by a small knob at the bend of one lead or by a dot on the package cover). Counting then proceeds through all the leads pointing in the same direction. When half the pins have been thus counted, the next pin in sequence is the one directly in line with the last one counted, but pointing in the opposite direction. Counting then proceeds across all the pins remaining until the highest-numbered pin is found directly opposite pin 1.

Two basic round forms of packaging also are manufactured. One of these is a cylinder topped with a dome and with a flat on the side of the cylinder that identifies the pin with the highest number. With the pins pointed toward you, they can be counted clockwise starting from the first pin beyond the flat.

The other round IC package is a "can" that resembles a pillbox hat with a narrow brim. A lug or tab on the rim identifies the pin with the highest number. With the pins pointed toward you, counting is done clockwise starting with pin 1 just beyond the lug (or tab).

As with DIP's, the round can packages can be used with sockets, or the leads can be pushed through drilled holes in a printed-circuit board before being soldered in place.

SEMICONDUCTORS 15.5

Since integrated circuits and many discrete semiconductor devices are now so common, some understanding of their operation is advisable. To this end we must consider again the atomic theory discussed in Section 1.6.

The previous discussion of atomic theory mentioned the number of electrons that may be found in the subshells of each shell of an atom. An important fact that was not mentioned is that the outermost atomic subshell never contains more than four pairs of electrons. Elements containing this maximum number are very resistive (insulators), but those with only one electron in the outer subshell are good conductors. Midway between these two extremes are elements called *semiconductors* (because they are neither good conductors nor good insulators). These have only two pairs of outermost electrons. Germanium and silicon fall into this category.

When silicon and germanium are very pure they form into extremely stable crystal structures through a process called **covalent bonding**. Since the outermost subshells of either material could accomodate another four electrons if the nucleus would support the additional charges, atoms cluster together so that each shares its space with an electron from each of four other atoms. Each electron is tightly held, therefore, and is not easily persuaded to leave its happy home and fly off as a free, conductive electron. In other words, the pure material is **intrinsically** of high resistivity (but not an insulator). Any impurity added to the material destroys the stability of the crystal and tends to increase the conductivity.

Two types of materials are commonly added to create special properties, but since they are specially selected and pure themselves, the word *impurity* is not really appropriate. To make this distinction, the preferred term is **dopant**, and the silicon or germanium modified by it is said to be **doped**.

If a few parts per million of an element with atoms possessing only three electrons in their outer subshells (**trivalent**) are diffused into an intrinsic crystal, an electron vacancy will occur at the location of each of these dopant atoms. The vacancy is called a **hole**. The crystal is then considered to be **acceptor** material because it will accept readily a free electron injected into it. Also, if an electric field is placed across the crystal, an electron from one of the atoms adjacent to the dopant may fill one hole but leave another hole in the atom from which it came. Such a process may repeat over and over with the result that the holes generally drift to the negative end of the crystal. Hole flow is thus the result of valence electrons moving from atom to adjacent atom toward the positive end of the crystal. Since doping with trivalent atoms causes the substance to be an acceptor of negatively-charged electrons, it behaves as though it were positively charged. Consequently, it is called P-type material.

Doping an intrinsic crystal with a few parts per million of an element containing five electrons in its outer subshell (**pentavalent**) causes one of the electrons to be "surplus" in the normal crystal structure. As a result, it is not tightly bound in orbit and can be dislodged easily to become a free electron. An electric potential placed across this type of crystal will cause these free electrons to flow quickly to the positive terminal. The crystal is said to be **donor** or *N*-type material.

SEMICONDUCTOR DIODES 15.6

If an intrinsic-semiconductor crystal is doped on one end with trivalent material and on the other end with pentavalent material the result is a diode. (See Figure 15.4.)

When a positive voltage is applied to the N-type material and a negative voltage is connected to the P-type, all the free electrons in the N-type material will rush to the positive terminal and any electrons that may have been present in the P-type material will be repelled from the negative charge. In this condition the diode is said to be *reverse biased*. The result can be seen in Figure 15.5. The main part of the crystal has been **depleted** (reduced in number) of **current carriers**. (Electrons carry current, of course, but so do holes. As electrons move from location to location leaving a hole and filling former holes, the hole has, seemingly, moved in the direction opposite to the electron movement.) The depleted area, therefore, has extremely high resistance.

On the other hand, if the polarity is changed on the diode, the current carriers are moved toward the junction between the N-type and P-type materials. This creates a sort of bridge that makes current flow easily. When the polarity is established in this way, the diode is said to be *forward biased*, as in Figure 15.6.

The diode works well as a rectifier of AC current. When the polarity of the voltage is in the forward-bias direction, current flows against a very low resistance. However, as soon as the polarity is reversed, the diode is reverse biased and the current must flow against very high resistance. The result is a pulsating direct-current flow.

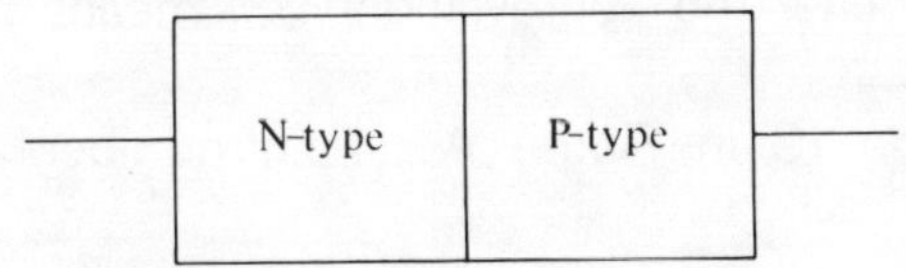

FIGURE 15.4 Diagram of a diode

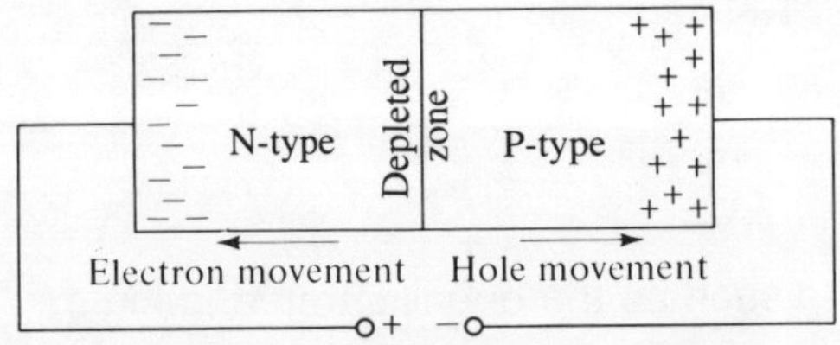

FIGURE 15.5 Diode reverse biased

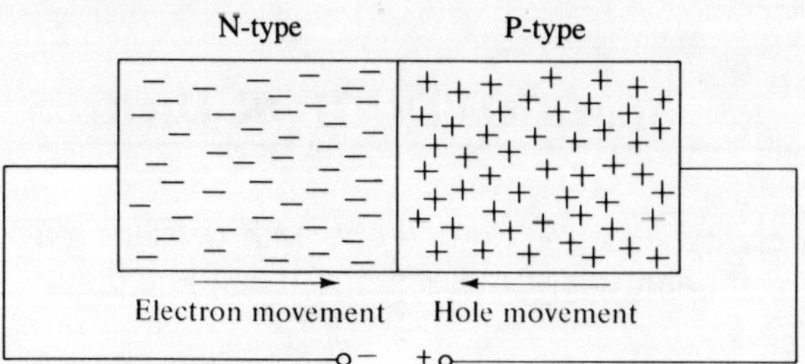

FIGURE 15.6   Diode forward biased

QUESTIONS

Select a number, word, or words from the following list to fill each of the blanks in the questions.

acceptor	hybrid	more	trivalent
chip	improved	neutrons	1
diode	in-line	package	2
discrete	integrated	pentavalent	7
donor	intrinsic	protons	8
dopant	junction	scale	10
dual	macro	section	14
electrons	medium	selected	24
emitter	merge	space	27
hole	modest	tetravalent	42

1. When speaking of integrated circuits, the abbreviation *MSI* means ________ ________ ________.

2. The last pin on the right of the IC in Figure A is pin number ________.

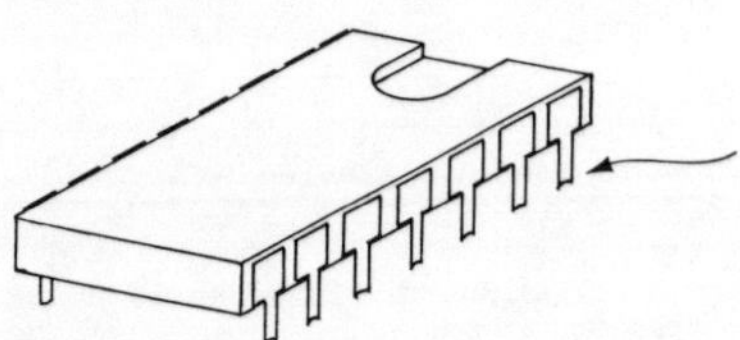

FIGURE A   DIP IC

3. The letters *DIP* as applied to IC's such as the one illustrated in Figure A stand for ________ ________ ________.

4. The vacant location in a semiconductor substance where an electron could fit into the crystal structure is called a __________.

5. Forward bias applied to a diode propels the current carriers toward the __________.

6. A semiconductor substance that is receptive of free electrons is called __________ material.

7. A circuit made up of separate and distinct units is said to be constructed of __________ parts.

8. When a small percentage of a semiconductor is a specially chosen foreign element, the additive is called a __________.

9. Extremely pure semiconductor elements are called __________ material.

10. An integrated circuit that contains two or more discrete parts is called a __________.

16

Amplifying Devices

The use of integrated circuits currently dominates most new electronic designs, but when an application requires the control of a large amount of power, tubes or transistors are usually employed. Discrete components have the physical size required for conducting large currents and can be manufactured with provisions for **heat sinking** (to conduct away excess heat). If the power requirements of a design are relatively modest, transistors are usually chosen as the active device. Sometimes, however, the special qualities of vacuum tubes are preferable, especially when high power is needed. For instance, all commercial broadcasting is still done with vacuum tubes because nothing else is available that can deliver the required power.

229

16.2 *BIPOLAR TRANSISTORS*

If a semiconductor crystal is doped with *N*-type material on both ends, and with *P*-type material in a narrow band in the center, the basic structure of a transistor is created. Figure 16.1 illustrates battery connections to these three sections (base, emitter, and collector) that will cause electrons to flow through the transistor from emitter to collector. Battery *a* causes the diode formed by the *N*-type and *P*-type material between *e* (emitter) and *b* (base) to be forward biased. Current will, therefore, flow easily. Because the base is lightly doped, the electrons from the emitter entering the base area to unite with holes will find very few to fill. Most of the electrons drifting in the *P*-type base material fall under the influence of the strong electric field provided by battery *d*. Consequently, they flow through to the collector, *c*, and then through resistor *R*. Actually, this collector current is more than 90 percent of all the charges entering the base from the emitter. In this way it is possible to produce a large voltage drop across *R* merely by controlling a small current *e* to *b*. This principle explains how the transistor illustrated can be used as an amplifier.

The foregoing explanation applies to an NPN transistor. If the crystal is made with *P*-type material on the ends and *N*-type material in the center, the transistor action is similar except that hole movement is substituted in the crystal for electron movement. In the connecting conductors, however, it is electrons that continue to be the current carriers. Also, the polarities applied to a PNP transistor are the reverse of those applied to an NPN for proper operation.

The symbols used for common transistors are shown in Figure 16.2. Note that the position of the arrow is the clue to whether the transistor is PNP or NPN.

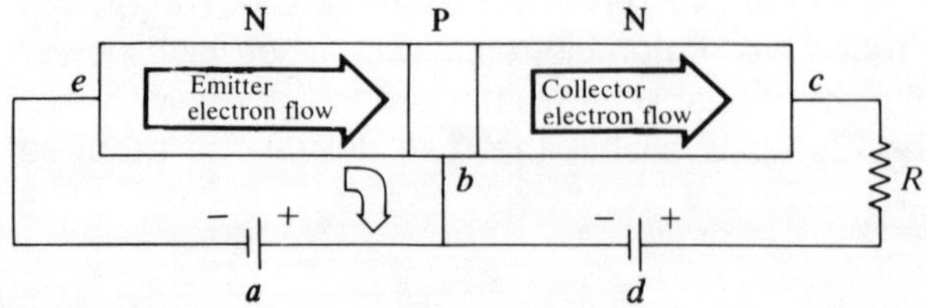

FIGURE 16.1 NPN transistor

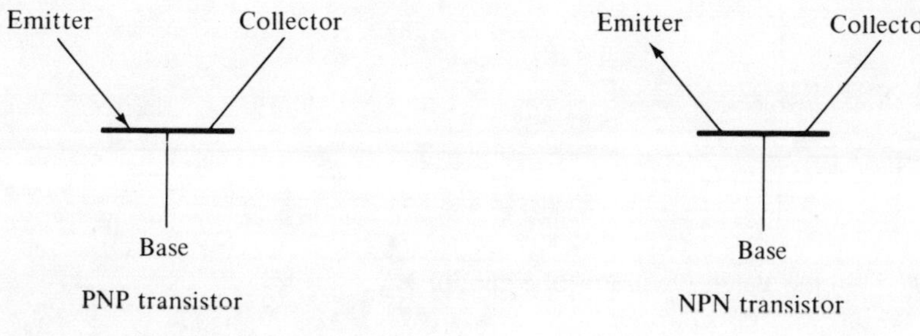

FIGURE 16.2 Transistor symbols

FIELD-EFFECT TRANSISTORS 16.3

Section 5.10 explained the basic kinds of FET's. Many of the integrated circuits available today are made with a type of FET. FET's have a number of advantages (and a few disadvantages) when compared with bipolar transistors. FET's have a much higher input impedance (which reduces loading to the signal source) and are basically voltage controlled, rather than current controlled like a bipolar device. Also, FET's are simpler and cheaper to mass-produce in IC's, and the CMOS (complementary metal oxide silicon) types consume very little power. However, insulated-gate FET's are so sensitive that the static electricity on a human hand can ruin them, and until recently FET's have been limited to low power—and relatively low frequency—applications. Now FET's are available that can operate at frequencies measured in gigahertz, or serve as the output of an audio power amplifier.

VACUUM TUBES 16.4

The most elementary vacuum tube is a diode that contains only an anode and a cathode. The diode was developed shortly after 1900 by Sir John Ambrose Fleming. Because it performs a function in electricity analogous to the one a check valve performs with fluids, it has often been called the **Fleming valve.** (See Figure 16.3.)

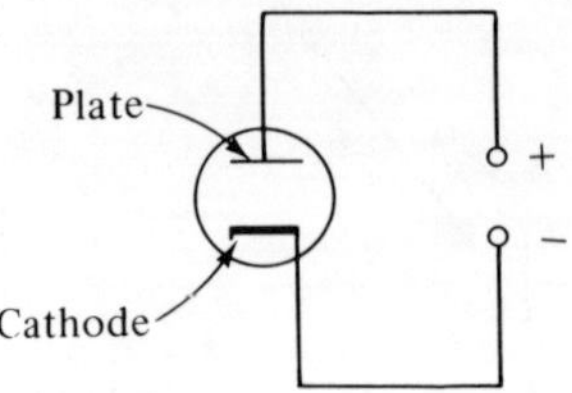

FIGURE 16.3 Fleming valve (vacuum-tube diode)

Within two years of Fleming's invention, Dr. Lee DeForest made another development that advanced the progress of the new technology of "wireless" (radio) communication spectacularly. A third element—the control grid—was inserted in the evacuated bulb between the cathode and anode to permit control of the internal resistance (or current flow) from cathode to anode. DeForest found that this enabled the amplification of electrical signals and named it the *Audion*. Today we call this device a **triode** and represent it by the schematic symbol in Figure 16.4.

Other elements have been added to vacuum tubes to improve certain characteristics, but these are refinements rather than a fundamental breakthrough. Of the vacuum tubes containing extra elements the most important types are the **tetrode** and the **pentode**. The tetrode has a second grid between the control grid and the anode that maintains a high positive charge even when a large current through the anode causes its supply voltage to drop low (due to attenuation through the load resistor). This **screen grid** increases the amplification possible with a tube and reduces capacitance between the control grid and the anode (which could cause oscillation). (See Figure 16.5.)

The pentode has a grid placed between the screen grid and the anode that is usually at ground potential. It is called a **suppressor grid** because it suppresses **secondary emission** of electrons from the

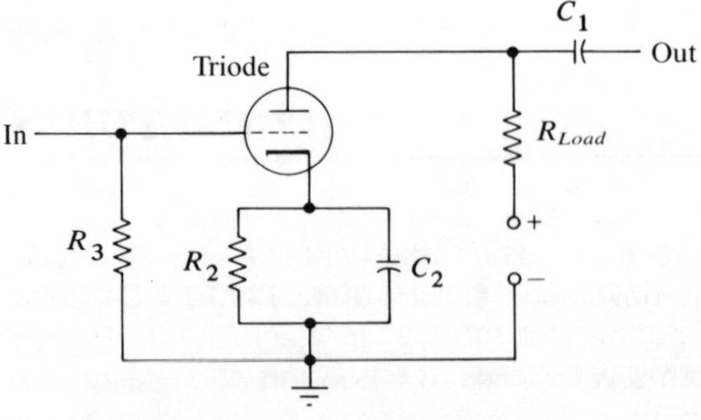

FIGURE 16.4 Triode amplifier

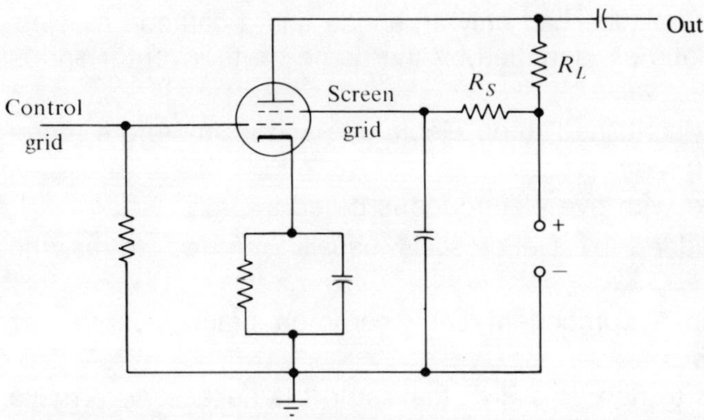

FIGURE 16.5 Tetrode amplifier

anode caused by the high temperatures and large currents made possible with the screen grid. The electrons are repelled back to the anode because the suppressor grid is negatively charged.

QUESTIONS

Select a letter, symbol, word, or words from the following list to fill each of the blanks in the questions.

capacitance	diode	lightly	power
cathode	Fleming	negative	screen
cold	frequency	NPN	suppressor
control	grid	pentode	tetrode
current	heat	PNP	triode
DeForest	heavily	positive	voltage

1. When there is proper forward bias to a transistor the collector current is nearly the same as the emitter current, because the base is very ___________ doped and provides little opportunity for the mating of current carriers.

2. If the arrow in the transistor symbol is pointed toward the base, the transistor is a(n) ___________.

3. Vacuum tubes are sometimes used in high-fidelty equipment because they produce fewer of certain types of distortion. Their use in broadcasting is due primarily to their capacity to handle high ___________.

4. Field-effect transistors are controlled by changes in ___________, but bipolar transistors are controlled by changes in ___________.

5. The original vacuum tube had only an anode and a cathode as electrodes. It is sometimes identified by the name of its inventor and is called a __________ valve.

6. The Audion tube contained three electrodes and is, therefore, commonly referred to as a __________.

7. The vacuum tube with five electrodes is called a __________.

8. The electrode added by DeForest is usually referred to as the __________ __________.

9. In order to keep a component cool, common practice is to use a __________ sink.

10. To cause current to flow in a PNP transistor it is necessary to make the emitter more __________ than the base.

17

Digital Circuits

The early chapters of this book investigated the operation of devices such as the public address system, radio, television, and the oscilloscope. In each instrument some kind of signal of varying amplitude or frequency was generated or amplified so that in the output the frequency was faithfully duplicated and the amplitude was increased proportionately. Such devices are classified as **analog** circuits.

Although all of the analog apparatus discussed are as important as ever, the greatest developments in electronics at the present time involve **digital** circuits. Such circuits are not only beginning to replace analog in the fields of television and recording, but also they are at the heart of development in **computers, microprocessors,** and **robotics.**

Digital circuits operate at two discrete (voltage) levels. One of these states is commonly at or very near zero voltage, and the other level at a significantly higher voltage. The two levels could be termed *off* and *on* as with a switch. In digital circuits, however, they are usually referred to as **zero** and **one.** With these two digits, zero and one, computers

235

do all of their calculations. At first it may seem impossible to do much with numbers when only two are available, but this situation is not the case once the **binary number system** is understood.

17.2 *COUNTING IN BINARY*

Before examining the binary number system, it is best to re-examine the common decimal system. To read a number, such as 275, it is understood that the 2 means 200, the 7 means 70, and the 5 means 5. The 2 became 200 because of its *position* in the third column from the right, which is termed the *hundreds column.* The 7 became 70 because it is in the second column, called the *tens column,* and the 5 remained itself because it is in the 1's column. Note that, from right to left, each succeeding column is ten times the value of the former, which is the reason for the name decimal.

In the binary system (moving from right to left) the first column is the 1's column, the second column is the 2's column, the third column represents the 4's, the fourth column represents 8's, and so on. Observe that with this system, each column has twice the value of the one on its right.

To write the decimal number 275 in binary, look at Table 17.1 and find a column value equal to or just under 275. That column number is 256, so write *1* for this ninth column. Next, look at the next column to the right to see if 128 more is needed to make 275. Since only 19 need be added to 256 to make 275, write a *0* for column eight. The 64 in the next column (moving right) is also too much, so again a *0* is written for this column. Writing *1* in the next column would add 32, which is also too much. Again, write *0*. The next column will add 16 and bring our total to 272, so place a *1* in that column. That leaves just 3 more to go, so *0*'s are written in the next two columns, and *1*'s are written in the last two. The binary number now looks like this: 100010011.

It is a wise discipline to avoid calling binary numbers by terms such as *ten*, *hundred*, and *thousand* for binary values such as 10, 100, or 1000. Table 17.2 compares decimal numbers with their binary equivalents and suggests the preferred method of verbally stating the latter.

TABLE 17.1 Position value in binary

2048	1024	512	256	128	64	32	16	8	4	2	1

TABLE 17.2   Counting in Binary

Decimal	Binary	(Verbal binary expression)
0	0	zero
1	1	one
2	10	one-zero
3	11	one-one
4	100	one-zero-zero
5	101	one-zero-one
6	110	one-one-zero
7	111	one-one-one
8	1000	one-zero-zero-zero
9	1001	one-zero-zero-one
10	1010	one-zero-one-zero
	etc.	

ADDING IN BINARY 17.3

Addition in binary arithmetic is practically the same as in decimal arithmetic. For example, add the binary expressions for 6 and 7.

$$6 = 0110$$
$$\underline{7 = 0111}$$
$$13 = 1101$$

In the first column add 0 and 1 to get 1. In the second column, 1 plus 1 is 10, so write down the 0 and carry the 1 to column three. In column three, 1 plus 1 equals 10, to which must be added the 1 carried over. Therefore, the total is 11. As in decimal arithmetic, write the 1 and carry 1. The fourth column is 0's, so the carried 1 is written. Note that the columns marked with a unit are the 8's column, the 4's column, and the 1's column: $8 + 4 + 1 = 13$.

BINARY-CODED DECIMAL 17.4

Sometimes, for convenience, numbers may be written in a form called **binary-coded decimal (BCD)**. A common example of this uses groups of four binary digits (**bits**) for each decimal digit. Each decimal digit then has the binary form shown in Table 17.2. The decimal number 275 then takes the following appearance:

$$0010 \quad 0111 \quad 0101$$
$$2 \qquad 7 \qquad 5$$

17.5 *OTHER NUMBER SYSTEMS USED*

All modern electronic calculators and computers use binary arithmetic internally to perform their calculations, but the input and output are usually presented in terms of other number systems. When a person uses a small calculator or programs a computer with a high-level language, it is normal to use the decimal number system. Programmers using **assembly language** usually use other systems. For instance, microprocessors are commonly programmed with **hexadecimal** numbers. This strange-sounding number system is based on 16. That is, the value of columns increases from right to left by a factor of 16, and there must be 16 digit symbols. Zero through nine are used as in the decimal system, but the succeeding six digits are *A* through *F*. Thus, *F* in the hexadecimal system is equivalent to decimal 15, decimal 16 becomes hexadecimal 10 (one–zero), decimal 17 becomes 11 (one–one), and so on.

17.6 *LOGIC MODULES*

In addition to learning how binary arithmetic works, anyone entering the digital field must learn about the various **logic modules** that make up the systems enabling a calculator or computer to do its job.

These modules are given strange sounding names such as **AND, OR, NAND, NOR, invertor,** etc. These are electronic circuits that can be assembled in various configurations to perform the desired arithmetic and control functions.

17.7 *AND GATE*

Recall from Section 17.1 that the two states of a digital circuit resemble the off and on states of a switch. In fact, an **AND gate** can be demon-

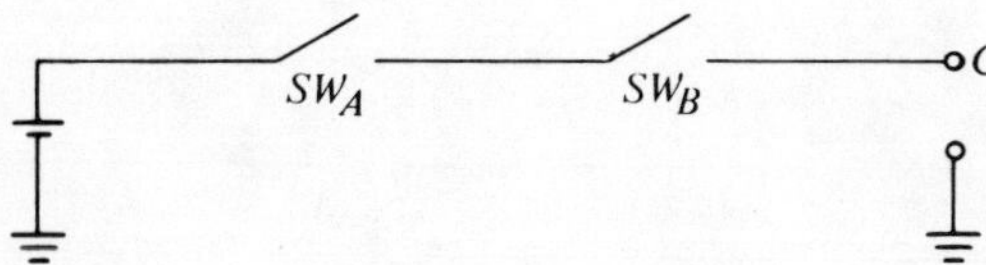

FIGURE 17.1 Simple AND gate

strated with two switches placed in series, as is illustrated in Figure 17.1.

Unless switch A (SW_A) *and* switch B(SW_B) are both closed, the voltage at C will be zero. In this ultra-simple case the closing of the switches is, of course, manual. In useful electronic modules it is necessary that this function be electronic. Commonly it is done with transistors. For planning a system, the symbol shown in Figure 17.2 is commonly used. In most cases today, rather than constructing the subsystem circuit from discrete components, an integrated circuit that performs the function is purchased.

A and B are inputs. If both A and B are high-voltage inputs, it is like closing switches 1 and 2 in Figure 17.1. The output C will be the high voltage. However, if *either* A or B is the low voltage, C will be low.*

OR GATE 17.8

As with the AND gate, an OR gate can be demonstrated with switches, as in Figure 17.3. If SW_A *or* SW_b, *or* both, are closed, the voltage at C will be high. The symbol of the OR gate is illustrated in Figure 17.4. If either A or B is high (positive system), C will be high.

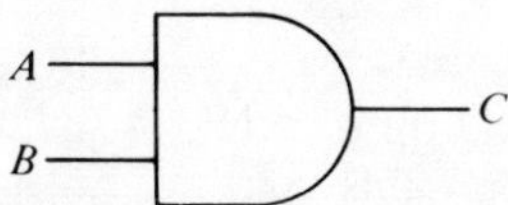

FIGURE 17.2 AND gate symbol

---

*Gates can also be used that operate in the reverse way. With an AND gate when A and B are both low, C will be low, but if either A or B is high, C is high. When this type of system is employed, it is said to be *negative logic*. The system described in the main text is positive logic.

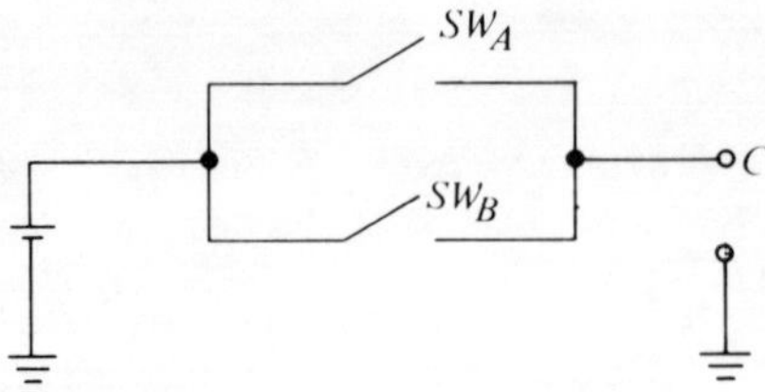

FIGURE 17.3   Simple OR gate

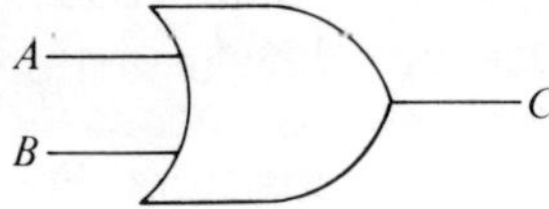

FIGURE 17.4   OR gate symbol

17.9 *INVERTOR/NOT*

It is often necessary to have a function that inverts an output. That is, if the logic of a circuit requires that a high voltage be converted to a low voltage and a low voltage changed to high voltage, we need an invertor. This function is not difficult to achieve. (A single stage of a common-emitter transistor amplifier causes inversion.) The symbol for an invertor is shown in Figure 17.5. This type is frequently called a NOT symbol because whatever the input is at X, the output at $\overline{X}$ is *not X* but is the opposite.

17.10 *NAND-NOR*

When the inversion function is combined with the AND gate, a NOT–AND (abbreviated NAND) gate is formed. Similarly when an OR gate is com-

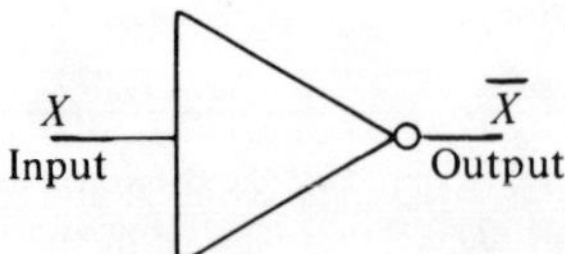

FIGURE 17.5   Invertor/NOT

bined with an invertor, a NOT–OR (NOR) gate results. Instead of the entire invertor symbol, the small circle (bubble) is applied to the AND and OR symbols to make the NAND and NOR symbols, as in Figure 17.6.

BOOLEAN ALGEBRA **17.11**

After World War II, when electronic computers were being developed, designers realized that the logic circuits they developed often were (unnecessarily) extremely complex, and subsequently could be simplified. Eventually they realized that a method had been developed one hundred years earlier for handling a binary logic problem by a specialized algebra. George Boole could scarcely have known how his **Boolean algebra** would be used. Today, logic is simplified on paper or by use of a computer to achieve the greatest economy of components before anything is wired.

Boolean algebra is a specialized study beyond the intent of this book. However, the term should be in the vocabulary of electronic technicians.

Another term associated with logic diagrams and Boolean algebra is **truth table.** For instance, the truth table for the OR gate (Figure 17.4) is shown here. Table 17.3 shows that when the inputs at A and B are low (0), the output at *C* is low. When either *A* or *B* is high (1), *C* is high.

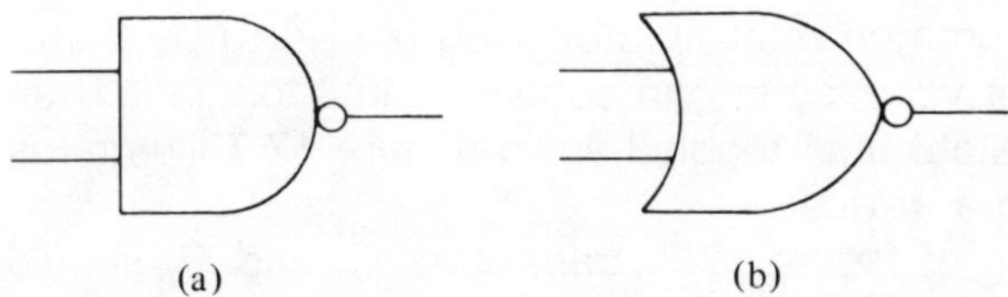

FIGURE 17.6 (a) NAND gate symbol; (b) NOR gate symbol

TABLE 17.3 Two input OR gate truth table

A	*B*	*C*
0	0	0
0	1	1
1	0	1
1	1	1

17.12 *MULTIVIBRATOR*

Three terms often used by digital technicians that usually baffle the uninitiated are **clock, toggle,** and **flip-flop.** A flip-flop is a circuit that effects the change of an output (from either a high or low state) to the opposite state by applying a short duration pulse (a high). As the name implies, a pulse "flips" the circuit to one state and another pulse "flops" it to the other state. If the low-voltage state in the output is considered a 0 and the high-voltage state is taken as a 1, an input pulse will flip a 0 to 1; the next input pulse will flop the 1 to 0, and so on. When each pulse arrives to perform its function, it is said to *toggle* the circuit. Perhaps the meaning of *toggle* will be easier to understand if you recognize that the kind of switch you commonly use to turn room lights on and off is called a toggle switch.

Gates, flip-flops, and invertors seemingly work instantaneously, but in reality they consume time in performing their functions. Even such short time intervals as 10 nanoseconds* per component can add up to an unbalanced condition in digital circuits. To cause circuits to operate in unison, it is usually necessary to use a timing device (**clock**) that puts out pulses at a steady frequency, enabling each flip-flop to be activated by its input only when the clock pulse is present. In other words, the clock keeps all flip-flops in step, like the drumbeat does for marchers.

Flip-flops and clock circuits commonly employ a configuration given the general name **multivibrator.** This circuit is really nothing more than a special type of oscillator that produces a square-shaped wave instead of a sine wave.

The most common multivibrator used is said to be **bistable.** This term means that it will stay in one condition until toggled. Then it will stay in the opposite state until toggled again. Figure 17.7 illustrates a simplified version of such a device.

In the circuit of Figure 17.7, transistors Q_1 and Q_2 should be very closely matched, but perfect matching never quite occurs. When V_{cc} is applied, one transistor will conduct better than the other. For this example, assume that it is Q_1. It will conduct because the base is connected by R_{B1} to the junction of R_{L2} and Q_2, which functions as a voltage divider and forward biases the Q_1 emitter-base junction. This bias causes Q_1 to **saturate** (conduct nearly like a short circuit). The result is that the junction of the Q_1 collector and R_{L1} are at nearly ground potential. This low potential reaches the base of Q_2 to place it at **cutoff**. Output Q_2 will then be at a maximum, and output Q_1 will be minimum. If a positive signal is

---

*This is about how long it takes for light to travel ten feet.

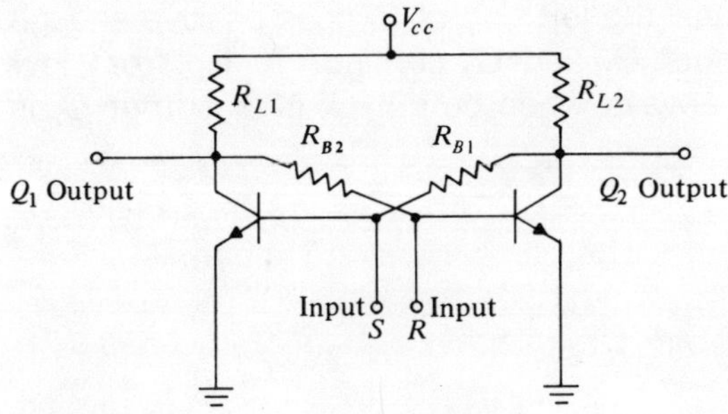

FIGURE 17.7 Bistable multivibrator

now brought in at input R, Q_2 will saturate, output Q_2 will fall to near ground potential, and Q_1 will go into cutoff. Output Q_1 will now be a maximum. This condition will continue to exist until a positive signal is placed at input S.

Clocks are most often **free-running** multivibrators. This means that the oscillator is inherently unstable and produces a constant frequency with an output of off, on, off, on (0, 1, 0, 1).

A simple free-running (**astable**) multivibrator makes use of a principle from Chapter 13. It was explained that when capacitors and resistors are placed in series, the time for charging and discharging the capacitor depends upon the RC time constant. In Figure 17.8, observe that R_1 and R_2 have been added to the circuit of Figure 17.7, while R_{B1} and R_{B2} have been replaced by C_1 and C_2.

When Q_1 fires (conducts), output Q_1 falls nearly to ground potential, forcing Q_2 into cutoff until the repelled electrons on the right side of C_2 can flow through R_2. Eventually the right side of C_2 achieves a positive charge from V_{cc}, and this forward biases Q_2 enough to cause it to fire. Output Q_2 then quickly drops to near ground potential, and Q_1 is driven

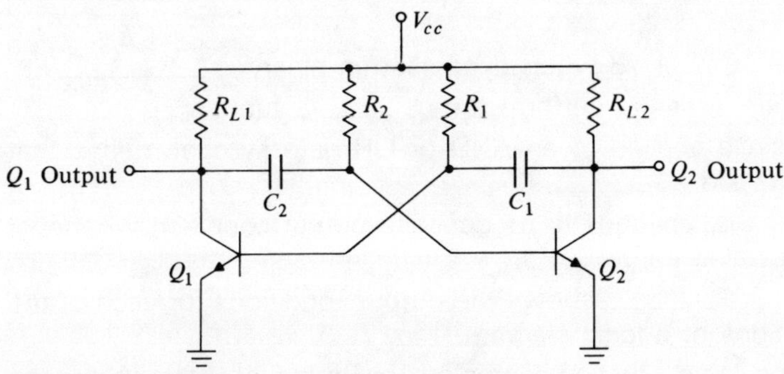

FIGURE 17.8 Free-running multivibrator

into cutoff. This condition exists until the left side of C_1 goes positive enough for Q_1 to fire again. Thus, outputs Q_1 and Q_2 continue to flip from low (binary 0) to high (binary 1) but always when output Q_1 is 0, output Q_2 is 1 and vice versa.

QUESTIONS

Select a number, symbol, word, or words from the following list to fill each of the blanks in the questions.

AND	NAND	0	32
bubble	NOR	1	56
flip-flop	NOT	2	77
inverted	OR	4	10010
logic	table	6	10101
multivibrator	toggle	8	110011
module	truth	21	1001001

1. The symbol shown in Figure A is an ___________ gate.

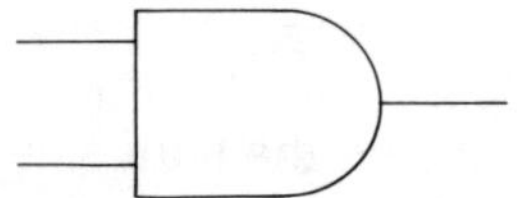

FIGURE A   Logic symbol

2. The binary number 1001101 when translated to decimal is ___________.

3. The decimal number 73 when expressed in binary is ___________.

4. Two switches in series perform the ___________ function.

5. The small circle applied to an AND or OR gate symbol means that the functions are ___________.

6. An oscillator that changes to its opposite output condition whenever triggered is called a ___________.

7. A ___________ ___________ shows the output conditions for each of the input conditions of a logic diagram.

8. To show the logic function, complete column f of Table A for the logic diagram in Figure B.

TABLE A

A	B	C	f
0	0	0	
0	0	1	
0	1	0	
0	1	1	
1	0	0	
1	0	1	
1	1	0	
1	1	1	

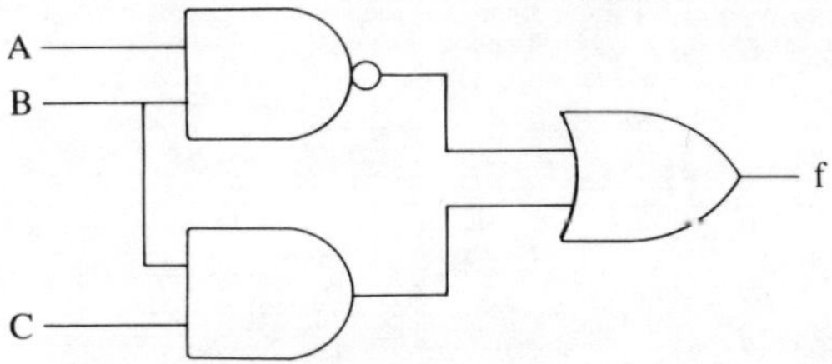

FIGURE B Logic diagram

9. The logic symbol in Figure C that represents a NAND gate is numbered __________.

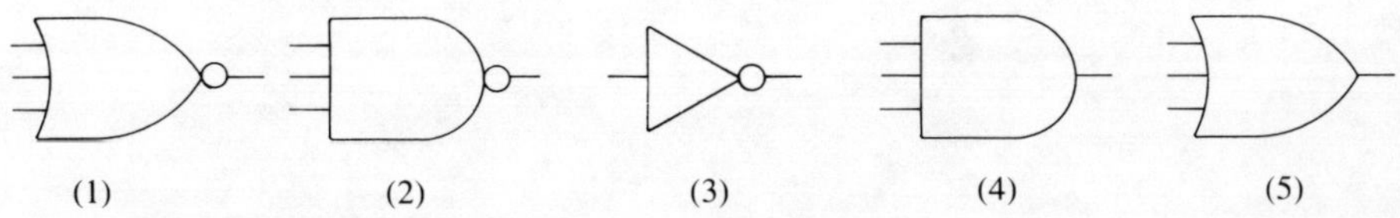

FIGURE C Logic symbols

10. The binary numbers 1010 and 1011, when added, total __________ in binary. This is the equivalent of decimal number __________.

18

Onward

SUMMARY 18.1

It has been the intention of the author to present some elementary areas in the field of electronics rather than an exhaustive and complete coverage of the subject. As you must now be aware, a comprehensive study of the field requires considerable use of mathematics. The more thoroughly one studies electronics, the greater is the requirement for skill in mathematical science. For a technician training in laboratory electronics at a community college, for instance, a good command of algebra and trigonometry is essential. If further study in a four-year college is attempted, a need for ability with calculus should be anticipated. If one expects to enter the field of computers, a knowledge of Boolean algebra should be acquired.

It will also be found to be most helpful to have a good general knowledge of physics. The application of electronics has now become so broad that a technician frequently finds a need for considerable knowledge in the fields where his electronic skills are being applied. In fact, many opportunities are developing in such diverse fields as bioelec-

247

tronics (biology/electronics), earthquake control (measurement of earth movement), ecology (measurements of many kinds, tracking of migration patterns, communication, and summations of data), disease diagnosis, mechanical mechanism performance monitoring and automatic adjustment, plus dozens more.

249